Alexander BOTTS

and the
EARTHWORM TRACTOR

BOTTS BEGINS

William Hazlett Upson

OCTANE
PRESS

Octane Press, Edition 1.0, September 2020
Volume 1
Copyright © 2020 by William Hazlett Upson
Alexander Botts illustrations © SEPS licensed by
Curtis Licensing Indianapolis, IN.

Main cover illustration by Hy Rubin
Cover portrait by Walter Skor
Interior illustrations by Tony Sarg

All rights reserved. With the exception of quoting brief passages for the purposes of review, no part of this publication may be reproduced without prior written permission from the publisher. The stories in this book were originally published in the *Saturday Evening Post*. While these stories have been reprinted in other books, this book is the only one to have the stories in their entirety as well as the original artwork.

ISBN: 978-1-64234-025-9
LCCN: 2020943737

Project Edited by Catherine Zinser
Design by Tom Heffron
Copyedited by Chelsey Clammer
Proofread by Peter Schletty

octanepress.com

Printed in Canada

CONTENTS

FOREWORD
By Lee Klancher
1

I'M A NATURAL-BORN SALESMAN
April 16, 1927
4

THE INDIRECT METHOD
August 6, 1927
22

I'M A HARD-BOILED BOZO
September 24, 1927
44

THE WONDERS OF SCIENCE
February 11, 1928
66

TROUBLE WITH THE EXPENSE ACCOUNT
February 25, 1928
86

THE BIG SALES TALK
April 7, 1928
110

SECONDHAND STUFF
September 8, 1928
130

THE BIG ENDURANCE TEST
September 22, 1928
150

BIG BUSINESS
October 13, 1928
168

SANDY INLET
December 15, 1928
188

THE OLD HOME TOWN
March 9, 1929
212

ALWAYS POLITE
March 30, 1929
228

ABOUT WILLIAM HAZLETT UPSON
252

ABOUT ALEXANDER BOTTS
253

FOREWORD

Conquering the Modern World with Mr. Alexander Botts

BY LEE KLANCHER

THE STORIES OF ALEXANDER BOTTS and his adventures selling Earthworm tractors were created during a time when machines were going to save the world. In 1927, when Botts was introduced to the world in the *Saturday Evening Post*, machines and factories were changing society at a breakneck pace. Motorized equipment was one of the wonders of the era, transforming people's daily lives and physically reshaping the world around them.

Things we take for granted today—a quick trip to the store that doesn't require hitching up a wagon or saddling a horse, bright lights in your home that turn on with the flip of a switch, and powerful machines capable of moving mountains—were headline news in the dynamic environment of the early 20th century.

In the stories penned by William Hazlett Upson, the Earthworm tractor is a modern marvel constantly performing previously unthinkable tasks. Impassable snow-covered roads are opened for traffic, vast housing tracts are cleared in days, forty acres of farmland are plowed in a single day, and the bottomless Great Gumbo Swamp that could not be crossed with horses and wagons is conquered with tracks and internal combustion.

While the wonders of new technology are a theme to the work, the central character, the bumbling master salesman Alexander Botts, was most likely the force who drove millions of readers to follow his escapades in *The Saturday Evening Post*. Comprised of letters and telegrams between Botts and the Earthworm Tractor Company, the stories inspired a popular movie, a short-lived comic strip, a one-act play, and nearly a dozen books.

Upson came to his topic area honestly. After trying his hand as a farmer and later serving time in an infantry division of the U.S. Army, Upson took a job working on the motor assembly line of the Caterpillar Tractor Company. He was pulled from the line and assigned to help the sales force put on field demonstrations of their crawlers. "My new job turned out to be about one-tenth demonstrating and nine-tenths repairing broken-down old tractors in cow sheds, barnyards, and on the open prairie," he wrote in a humorous autobiographical piece, "Ergophobia," published in the June 24, 1933, edition of *The Saturday Evening Post*.

Upson's work with Caterpillar led him to his true calling. "The tractors themselves are truly fascinating pieces of machinery. The people who own them and work with them are, almost without exception, splendid fellows. And the business is so full of adventures and idiotic incidents that it was a delight to be associated with it."

Believing that the general public should know more about the machines and people who ran them, Upson began writing the Alexander Botts stories. Once *The Saturday Evening Post* picked them up, they became quite popular and Upson had found himself a profession writing about one of the newest, most exciting technologies of his time.

Upson's experiences gave him a well of authenticity to his machine stories, but there is more to them than just heavy equipment saving the day. Upson's entertaining prose paints Botts as a bit of a self-aggrandizing buffoon willing to pull off stunts to garner a sale, and certainly part of what makes the stories fun to read is seeing through Bott's grandiose declarations to the foolishness that actually took place.

Underneath the satire and salesmanship of Upson's character lies the solid ground of core American values. While Botts is willing to stretch the rules to make a sale, the victim always benefits from the wonders of the new tracked machines.

And as we get to know Botts, his bedrock values extend well beyond the entertaining shell of an inexperienced salesman unwilling to own up to his considerable mistakes. As the stories progress and Upson shapes his central character, Botts emerges as a bit of an American working class hero, or at the least a figure whose often misguided actions lead to end results that benefit the honest and the underprivileged.

FOREWORD

When Botts is sent out to repossess a machine from a young couple unable to make their payments, he bumbles his way into finding them prosperity. A crooked politician is dispatched, as are multiple other figures of questionable character. The underdogs are served by the blundering salesman, who even develops a little humility in "The Big Sales Talk."

Upson's craft improves during the course of the stories as well. Not only does his prose get more colorful, Upson also concocts increasingly outrageous and cleverly structured situations for Botts and the Earthworm tractors to confront.

Those are the big draws, the literary substance that satisfies the soul. The details are equally appealing. Botts is on the road during an era when communication to the head office was akin to radio communication with the moon. His notes are sent via mail and telegram, and Botts has no choice but to work independently. That is part of the amusement, because so many of his decisions lead to outrageous situations.

The distance also makes his relationship with his employer tumultuous, and he as often as not gets fired, or quits. Somehow, through all this, the ever upbeat Botts makes the sale and keeps his job.

Lastly, part of the joy of the book that rings like a bell today is traveling back to remote parts of our world from nearly 100 years ago. On the road, far from the cities of then and the bothersome connectivity of today, Botts roams free in a world where snow closes roads for weeks on end, visitors are put up as guests without question, microphones are magical devices, and the most vexing problem of the day can be solved with a sixty-horsepower crawler.

I'M A NATURAL-BORN SALESMAN

ILLUSTRATED BY TONY SARG

STONEWALL JACKSON HOTEL,
MEMPHIS, TENNESSEE.
MARCH 11, 1920.

THE FARMERS' FRIEND TRACTOR COMPANY,
EARTHWORM CITY, ILLINOIS.

GENTLEMEN: I have decided you are the best tractor company in the country, and consequently I am giving you first chance to hire me as your salesman to sell tractors in this region.

I'm a natural-born salesman, have a very quick mind, am twenty-eight years old, am honest and reliable and can give references if required. I have already had considerable experience as a machinery salesman, and I became familiar with your Earthworm tractors as a member of the motorized field artillery in France. I can demonstrate tractors as well as sell them.

When do I start work?

Very truly yours,
ALEXANDER BOTTS.

FARMERS' FRIEND TRACTOR COMPANY
MAKERS OF EARTHWORM TRACTORS
EARTHWORM CITY, ILLINOIS

MR. ALEXANDER BOTTS,   MARCH 13, 1920.
STONEWALL JACKSON HOTEL,
MEMPHIS, TENNESSEE.

DEAR MR. BOTTS: Your letter is received. We have no opening for a salesman at present, but we are badly in need of a service mechanic. As you say you are familiar with our tractors, we will try you out on this job at one hundred dollars per month plus traveling expenses.

You will report at once to our Mr. George Healy, salesman for Tennessee and Mississippi, who is now at the Dartmouth Hotel, Memphis. You will go with him to Cyprus City, Mississippi, to demonstrate a ten-ton

Earthworm tractor for Mr. Jackson, a lumber operator of that place. Mr. Healy will tell you just what you are to do.

We enclose a check for one hundred dollars advance expense money.

> Very truly,
> GILBERT HENDERSON,
> *Sales Manager.*

> STONEWALL JACKSON HOTEL,
> MEMPHIS, TENNESSEE.
> MARCH 16, 1920.

THE FARMERS' FRIEND TRACTOR COMPANY,
EARTHWORM CITY, ILLINOIS.

GENTLEMEN: As soon as your letter came, I went around to see Mr. Healy, and it is lucky for you that you hired me, because Mr. Healy has just been taken sick with appendicitis. They were getting ready to take him to the hospital, and he was pretty weak, but he managed to tell me that the tractor for the demonstration had already arrived at the freight station in Cyprus City.

He also explained that this Mr. Jackson down there owns about a million feet of cypress timber which he wants to get out and sell right away before the present high price of lumber goes down. It seems the ground is so swampy and soft from the winter rains that with his present equipment of mules and wagons he won't be able to move any of his timber until summer.

But Mr. Healy was down there a couple of weeks ago, and he arranged to put on a demonstration to show Mr. Jackson that an Earthworm tractor can go into those swamps and drag out the timber right away. Mr. Jackson said he would buy the tractor if it did the work, and Mr. Healy was feeling very low because he was sick and couldn't go down to hold the demonstration.

"You can rest easy, Mr. Healy," I said. "When you look at me, you're gazing on a natural-born salesman. I will go down there and do your work, as well as mine. I will put on a swell demonstration, and then I will sell the goods."

As Mr. Healy did not seem to know just what to say to this, I gathered up all his order blanks, selling literature, price lists, and so on, and also the bill of lading and the check to pay the freight on the tractor. Then I wished him good luck and left.

From this you can see that I am quick to grasp an opportunity, and that you made no mistake in hiring me. I am leaving for Cyprus City tonight.

Cordially yours,
ALEXANDER BOTTS.

FARMERS' FRIEND TRACTOR COMPANY
SALESMAN'S DAILY REPORT

DATE: MARCH 17, 1920.
WRITTEN FROM: DELTA HOTEL, CYPRUS CITY, MISSISSIPPI.
WRITTEN BY: ALEXANDER BOTTS, SERVICE MECHANIC AND PINCH HITTER SALESMAN.

I found this pad of salesman's report blanks among the stuff I got from Mr. Healy. I see by the instructions on the cover that each salesman is supposed to send in a full and complete report of everything he does, so I will give you all particulars of a very busy day.

I arrived at 7:51 this morning at Cyprus City—which turns out to be pretty much a country town in what they call the Yazoo Delta. The whole country here is nothing but a swamp, and the main street of the town ends in a high bank that they call a levee, on the other side of which is the Mississippi River flowing along about twenty feet higher than the town.

After alighting from the train, and after noting that it was a cloudy day and looked like rain, I engaged a room at the Delta Hotel. I then hurried over to the freight station, where I found the big ten-ton Earthworm tractor on the unloading platform. They had dragged it off the car with a block and tackle. And when I saw that beautiful machine standing there so big and powerful, with its fine wide tracks like an army tank, with its elegant new shiny paint and with its stylish cab for the driver, I will admit that I felt a glow of pride to think that I was the salesman and service mechanic for such a splendid piece of machinery.

Note: Of course, as I said in my letter, I am an old machinery salesman. But the largest thing I ever sold before was the Excelsior Peerless Self Adjusting Automatic Safety Razorblade Sharpener. I did very well with this machine, but I could not take the pride in it that I feel I am going to have in this wonderful ten-ton Earthworm tractor.

After paying the freight I hired several guys from the town garage to put gas and oil in the tractor, and then I started them bolting the little cleats onto the tracks. You see I am right up on my toes all the time. I think of everything. And I figured that if we were going through the mud we would need these cleats to prevent slipping. While they were being put on, I stepped over to the office of Mr. Johnson, the lumberman.

Note: This bird's name is Johnson—not Jackson, as you and Mr. Healy told me. Also, it strikes me that Mr. Healy may have been fairly sick even as long as two weeks ago, when he was down here. In addition to getting the name wrong he did very poor work in preparing this prospect. He did not seem to be in a buying mood at all.

As soon as I had explained my errand to this Mr. Johnson—who is a very large, hard-boiled bozo—he gave me what you might call a horse laugh.

"You are wasting your time," he said. "I told that fool salesman who was here before that tractors would be no good to me. All my timber is four miles away on the other side of the Great Gumbo Swamp, which means that it would have to be brought through mud that is deeper and stickier than anything you ever seen, young feller."

"You would like to get it out, wouldn't you?" I asked.

"I sure would," he said, "but it's impossible. You don't understand conditions down here. Right on the roads the mules and horses sink in up to their bellies; and when you get off the roads, even ducks and turtles can hardly navigate."

"The Earthworm tractor," I said, "has more power than any duck or turtle. And if you'll come out with me, I'll show you that I can pull your logs through that swamp."

"I can't afford to waste my time with such crazy ideas," he said. "I've tried motor equipment. I have a motor truck now that is stuck three feet deep right on the main road at the edge of town."

"All right," I said, always quick to grasp an opportunity, "how about coming along with me while I pull out your truck?"

"Well," said Mr. Johnson, "I can spare about an hour this morning. If you'll go right now, I'll go with you, although I doubt if you can even pull out the truck. And even if you do, I won't buy your tractor."

"How about going this afternoon?" I asked.

"I'll be busy every minute of this afternoon. It's now or never."

"Come on!" I said.

We walked over together to the freight platform, and as the cleats were now all bolted on we both climbed into the cab.

Note: I will explain that I was sorry that Mr. Johnson had been unable to wait until afternoon, as I had intended to use the morning in practicing up on driving the machine. It is true, as I said in my letter, that I became familiar with Earthworm tractors when I was a member of a motorized artillery outfit in France, but as my job in the artillery was that of cook, and as I had never before sat in the seat of one of these tractors, I was not as familiar with the details of driving as I might have wished. However, I was pleased to see that the tractor seemed to have a clutch and gear shift like the automobiles I have often driven.

I sat down on the driver's seat with reasonable confidence, Mr. Johnson sat down beside me, and one of the garage men cranked up the motor. It started at once, and when I heard the splendid roar of the powerful exhaust, and saw that thirty or forty of the inhabitants were standing around with wondering and admiring faces, I can tell you I felt proud of myself. I put the gear in low, opened the throttle and let in the clutch.

Note: I would suggest that you tell your chief engineer, or whoever it is that designs your tractors, that he ought to put in a standard gear shift. You can understand that it is very annoying, after you have pulled the gearshift lever to the left and then back, to find that instead of being in low you are really in reverse.

As I said, I opened the throttle, let in the clutch, and started forward. But I found that when I started forward, I was really—on account of the funny gear shift—moving backward. And instead of going down the gentle slope of the ramp in front, the whole works backed off the rear edge of the platform, dropping at least four feet into a pile of crates with such a sickening crash that I thought the machine was wrecked and both of us killed.

But it soon appeared that, although we were both very much shaken up, we were still alive—especially Mr. Johnson, who began talking so loud and vigorously that I saw I need have no worry about his health. After I had got Mr. Johnson quieted down a bit, I inspected the machine and found that it was not hurt at all. As I am always alert to seize an opportunity, I told Mr. Johnson that I had run off the platform on purpose to show him how strongly built the tractor was. Then after I had promised I would not make any more of these jumps, he consented to remain in the tractor, and we started off again.

Note: Kindly tell your chief engineer that Alexander Botts congratulates him on producing a practically unbreakable tractor. But tell him that I wish he would design some thicker and softer seat cushions. If the base of the chief engineer's spine was as sore as mine still is, he would realize that there are times when good thick seat cushions are highly desirable.

As we drove up the main street of Cyprus City, with a large crowd of admiring townsfolk following after, I seemed to smell something burning. At once I stopped, opened up the hood and discovered that the paint on the cylinders was crackling and smoking like bacon in a frying pan.

"Perhaps," suggested Mr. Johnson, "there is no water in the radiator."

I promptly inspected the radiator, and, sure enough, that was the trouble.

Note: I would suggest that if your chief engineer would design an air-cooled motor for the tractor, such incidents as the above would be avoided.

I borrowed a pail from a store and filled the radiator. Apparently, owing to my alertness in this emergency, no damage had been done.

When we started up again, we had not gone more than a few yards before I felt the tractor give a little lurch. After we had got a little farther along, I looked back, and right at the side of the street I saw one of the biggest fountains I have ever seen in all my life. A solid column of water about eight inches thick was spouting high in the air, spreading out at the top like a mushroom, and raining down all around like Niagara Falls.

"I can't afford to waste my time with such crazy ideas"

I heard somebody yell something about a fire plug; and as I have a quick mind, I saw right away what had happened. The hood of the tractor is so big that it had prevented me from seeing a fire plug right in front of me. I had unfortunately run into it, and as it was of very cheap, inferior construction, it had broken right off.

For a while there was great excitement, with people running here and there, hollering and yelling. The sheriff came up and took my name, as he seemed to think I was to blame—in spite of the fact that the fire plug was in such an exposed position. I was a bit worried at the way the water was accumulating in the street, and consequently I was much relieved when they finally got hold of the waterworks authorities and got the fountain turned off. You see the fire mains here are connected to the Mississippi River, and if they had not turned the water off the whole river would have flowed into the business district of Cyprus City.

Note: I would suggest that your chief engineer design these tractor hoods a little lower, so as to avoid such accidents in the future.

After the water had been turned off, we got under way again, clanking along the main street in high gear and then driving out of town to the eastward over one of the muddiest roads I ever saw. The tractor, on account of its wide tracks, stayed right up on top of the mud and rolled along as easy and smooth as a Pullman car. Behind us a large crowd of local sightseers floundered along as best they could; some of them wading through the mud and slop and others riding in buggies pulled by horses or mules.

Mr. Johnson acted as if he was pretty sore, and I did not blame him. Although the various mishaps and accidents we had been through were unavoidable and not my fault at all, I could understand that they might have been very annoying to my passenger. Perhaps that is one reason I am such a good salesman; I can always get the other feller's point of view. I livened up the journey a bit by telling Mr. Johnson a number of Irish jokes, but I did not seem to get any laughs, possibly because the motor made so much noise Mr. Johnson couldn't hear me.

By this time, I had got the hang of driving the machine very well, and I was going along like a veteran. When we reached Mr. Johnson's truck—which was deep in the mud at the side of the road about half a mile from town—I swung around and backed up in front of it in great style.

The road, as I have said, was soft and muddy enough; but off to the right was a low flat stretch of swamp land that looked much muddier and a whole lot softer. There were patches of standing water here and there, and most of it was covered with canebrake—which is a growth of tall canes that look like bamboo fishing poles.

Mr. Johnson pointed out over this mass of canebrake and mud. "That is an arm of the Great Gumbo Swamp," he yelled very loud, so I could hear him above the noise of the motor. "Your machine may be able to navigate these roads, but it would never pull a load through a slough like that."

I rather doubted it myself, but I didn't admit it. "First of all," I said, "we'll pull out this truck."

We both got out of the tractor, and right away we sank up to our knees in the soft, sticky mud. The truck was a big one, loaded with lumber, and it was mired down so deep that the wheels were practically out of sight, and the body seemed to be resting on the ground. Mr. Johnson didn't think the tractor could budge it, but I told him to get into the driver's seat of the truck so he could steer it when it got going.

By this time a gentle rain had started up, and Mr. Johnson told me to hurry up, as the truck had no cab and he was getting wet. I grabbed a big chain out of the truck toolbox, and told Mr. Johnson to get out his watch. He did so.

"In just thirty seconds," I said, "things are going to start moving around here."

I then rapidly hooked one end of the chain to the back of the tractor, fastened the other end to the truck, sprang into the tractor seat and started the splendid machine moving forward. As the tractor rolled steadily and powerfully down the road, I could hear the shouting of the crowd even above the noise of the motor. Looking around, however, I saw that something was wrong. The truck—or rather, the major portion of it—was still in the same place, and I was pulling only the radiator. As I have a quick mind I saw at once what had happened. Quite naturally I had slung the chain around the handiest thing on the front of the truck—which happened to be the radiator cap. And as the truck was of a cheap make, with the radiator not properly anchored, it had come off.

I stopped at once, and then I had to spend about ten minutes calming down Mr. Johnson by assuring him that the Farmers' Friend Tractor Company would pay for a new radiator. I backed up to the truck again, and Mr. Johnson took the chain himself, and by burrowing down in the mud managed to get it fastened around the front axle. Then he climbed back into the seat of the truck and scowled at me very disagreeably. By this time the rain was falling fairly briskly, and this may have had something to do with his ill humor.

When I started up again everything went well. The motor roared, the cleats on the tracks dug into the mud, and slowly and majestically the tractor moved down the road, dragging the heavy truck through the mud behind it.

At this point I stuck my head out of the tractor cab to acknowledge the cheers of the bystanders, and in so doing I unfortunately knocked off my hat, which was caught by the wind and blown some distance away. At once I jumped out and began chasing it through the mud. The crowd began to shout and yell, but I paid no attention to this noise until I had reached my hat and picked it up—which took me some time, as the hat had blown a good ways, and I could not make any speed through the mud. When at last I looked around I saw that a curious thing had happened.

In getting out of the tractor I had accidentally pulled on one of the handlebars enough to turn the tractor sidewise. And in my natural excitement—the hat having cost me $8.98 last week in Memphis—I had forgotten to pull out the clutch. So when I looked up I saw that the tractor, with Mr. Johnson and his truck in tow, was headed right out into the Great Gumbo Swamp. It had already got a good start and it was going strong. As Mr. Johnson seemed to be waving and yelling for help, I ran after him.

At once I jumped out and began chasing it through the mud.

But as soon as I got off the road the mud was so deep and soft that I could make no headway at all. Several of the bystanders also attempted to follow, but had to give it up as a bad job. There was nothing to do but let poor Mr. Johnson go dragging off through the swamp.

Although I was really sorry to see Mr. Johnson going off all by himself this way, with no protection from the pouring rain, I could not help feeling a thrill of pride when I saw how the great ten-ton Earthworm tractor was eating up that terrible soft mud.

The wide tracks kept it from sinking in more than a few inches; the cleats gave it good traction; and the motor was so powerful that it pulled that big truck like it was a mere match box, and this in spite of the fact that the truck sank in so deep that it plowed a regular ditch as it went along.

As I am a natural-born salesman and quick to grasp every opportunity, I yelled a little sales talk after Mr. Johnson. "It's all right!" I hollered. "I'm doing this on purpose to show you that the Earthworm can go through any swamp you got!" But I doubt if he heard me; the roar of the tractor motor was too loud. And a moment later the tractor, the truck and Mr. Johnson had disappeared in the canebrake.

While I was considering what to do next, a nice looking man in a corduroy suit came over to me from one of the groups of bystanders. "This is only an arm of the Great Gumbo Swamp," he said. "If that tractor doesn't mire down, and if it goes straight, it will come out on the levee on the other side, about a mile from here."

"An Earthworm tractor never mires down," I said. "And as long as there is nobody there to pull on the handlebars, it can't help going straight."

"All right," said the man, "If you want to hop in my buggy, I'll drive you back to town and out the levee so we can meet it when it gets there."

"Fine!" I said. "Let's go." I have always been noted for my quick decisions, being similar to Napoleon in this particular. I at once climbed in the buggy with the man in the corduroy suit, and he drove the horse as fast as possible into town and then out the levee, with all the sightseers plowing along behind, both on foot and in buggies.

When we reached the place where the tractor ought to come out, we stopped and listened. Far out in the swamp we could hear the roar of the tractor motor. It got gradually louder and louder. We waited. It was still raining hard. Suddenly there was a shout from the crowd. The tractor came nosing out of the canebrake, and a moment later it had reached the bottom of the levee, with the big truck and Mr. Johnson dragging along behind. As the tractor was in low gear, I had no trouble in jumping aboard and stopping it; and it is just as well I was there to do this. If I had not stopped it, it would have shot right on over the levee and into the Mississippi River, probably drowning poor Mr. Johnson.

As it was, Mr. Johnson was as wet as a sponge, on account of the heavy rain, and because he had been too cheap to get himself a truck with a cab on it. But he was a long way from being drowned. In fact, he seemed very lively; and as I got down from the tractor he jumped out of the truck and came running at me, waving his arms around, and shouting and yelling, and with a very dirty look on his face. What he had to say to me would fill a small book; in fact, he said so much that I'm afraid I will have to put off telling you about it until my report tomorrow.

It is now midnight and I am very tired, so I will merely enclose my expense account for the day and wish you a pleasant good night. Kindly send check to cover expenses as soon as possible. As you will see, my $100 advance is already gone, and I have had to pay money out of my own pocket.

Cordially yours,
ALEXANDER BOTTS.

EXPENSE ACCOUNT

Railroad fare (Memphis to Cyprus City)	$ 6.10
Pullman ticket	3.20
Gas and oil for tractor	8.50
Labor (putting on cleats, and so on)	9.00
36 doz. eggs @ 50c per doz.	18.00

    NOTE: It seems the crates we landed on when we dropped off the freight platform were full of eggs.

1 Plate glass window	80.00

    NOTE: I forgot to say in my report that in the confusion following the break of the fire plug I accidentally sideswiped a drug store with the tractor.

Radiator for truck, and labor to install	46.75
Cleaning hat and pressing trousers	3.50
Total	$175.05

    NOTE: I will list the hotel bill, the bill for the fire plug and other expenses when I pay them.

Farmers' Friend Tractor Company
Salesman's Daily Report

Date: March 18, 1920.
Written from: Delta Hotel, Cyprus City, Mississippi.
Written by: Alexander Botts.

I will take up the report of my activities at the point where I stopped yesterday when Mr. Johnson had just gotten out of the truck and was coming in my direction. As I stated, he had a good deal to say. Instead of being grateful to me for having given him such a splendid demonstration of the ability of the Earthworm tractor to go through a swamp, and instead of thanking me for saving his life by stopping him just as he was about to shoot over the levee into the Mississippi River, he began using very abusive language, which I will not repeat, except to say that he told me he would not buy my tractor, and that he never wanted to see me or my damn machinery again. He also said he was going to slam me down in the mud and jump on my face, and it took six of the bystanders to hold him and prevent him from doing this. And, although there were six of them, they had a lot of trouble holding him, owing to the fact that he was so wet and slippery from the rain.

As I am a natural-born salesman I saw right away that this was not an auspicious time to give Mr. Johnson any sales talk about tractors. I decided to wait until later, and I walked back to the tractor in a dignified manner, looking back over my shoulder, however, to make sure Mr. Johnson was not getting away from the guys that were holding him.

After they had led Mr. Johnson back to town I made up my mind to be a good sport, and I hauled his truck into town and left it at the garage to be repaired. The rest of the day I spent settling up various expense items—which appeared on my yesterday's expense account—and in writing up my report. When I finally went to bed at midnight, it was with a glow of pride that I thought of the splendid work I had done on the first day of my employment with the great Farmers' Friend Tractor Company, Makers of Earthworm Tractors. Although I had not as yet made any sales, I could congratulate myself on having put on the best tractor demonstration ever seen in Cyprus City, Mississippi.

This morning, after breakfast, I had a visit from the nice looking man in the corduroy suit who gave me the buggy ride yesterday.

"I am a lumber operator," he said, "and I have a lot of cypress back in the swamps that I have been wanting to get out. I haven't been able to move it because the ground has been so soft. However, since I saw your tractor drag that big heavy truck through the swamp yesterday, I know that it is just what I want, I understand the price is $6000, and if you will let me have the machine right away I will take you over to the bank and give you a certified check for that amount."

"Well," I said, "I was supposed to sell this machine to Mr. Johnson, but as he has had a chance at it and hasn't taken it, I suppose I might as well let you have it."

"I don't see why you gave him first chance," said the man in the corduroy suit. "When your other salesman, Mr. Healy, was down here, I gave him more encouragement than anybody else he talked to. And he said he would ship a tractor down here and put on a demonstration for me."

"What is your name?" I asked.

"William Jackson," he said.

As I have a quick mind I saw at once what had happened. This was the guy I had been supposed to give the demonstration for in the first place, but I had very naturally confused his name with that of Mr. Johnson. There ought to be a law against two men with similar names being in the same business in the same town.

However, it had come out all right. And as I am a natural-born salesman, I decided that the thing to do was to take Mr. Jackson over to the bank right away—which I did. And now the tractor is his.

I enclose the certified check. And I have decided to remain in town several days more on the chance of selling some more machines.

Cordially yours,
ALEXANDER BOTTS.

TELEGRAM
EARTHWORM CITY ILL 1015A MAR 19 1920
ALEXANDER BOTTS
DELTA HOTEL
CYPRUS CITY MISS

YOUR FIRST REPORT AND EXPENSE ACCOUNT
RECEIVED STOP YOU ARE FIRED STOP WILL DISCUSS
THAT EXPENSE ACCOUNT BY LETTER STOP IF YOU SO
MUCH AS TOUCH THAT TRACTOR AGAIN WE WILL
PROSECUTE YOU TO THE FULLEST EXTENT OF THE LAW

     FARMERS FRIEND TRACTOR COMPANY
     GILBERT HENDERSON SALES MANAGER

     NIGHT LETTER
CYPRUS CITY MISS 510P MAR 19 1920
FARMERS FRIEND TRACTOR CO
EARTHWORM CITY ILL

YOUR TELEGRAM HERE STOP WAIT TILL YOU GET MY
SECOND REPORT STOP AND THAT IS NOT ALL STOP THE
WHOLE TOWN IS TALKING ABOUT MY WONDERFUL
TRACTOR DEMONSTRATION STOP JOHNSON HAS
COME AROUND AND ORDERED TWO TRACTORS
STOP THE LEVEE CONSTRUCTION COMPANY OF THIS
PLACE HAS ORDERED ONE STOP NEXT WEEK IS TO
BE QUOTE USE MORE TRACTORS WEEK UNQUOTE
IN CYPRUS CITY STOP MASS MEETING MONDAY TO
DECIDE HOW MANY EARTHWORMS THE CITY WILL
BUY FOR GRADING ROADS STOP LUMBERMENS MASS
MEETING TUESDAY AT WHICH I WILL URGE THEM TO
BUY TRACTORS AND JACKSON AND JOHNSON WILL

BACK ME UP STOP WEDNESDAY THURSDAY FRIDAY AND SATURDAY RESERVED FOR WRITING UP ORDERS FROM LUMBERMEN CONTRACTORS AND OTHERS STOP TELL YOUR CHIEF ENGINEER TO GET READY TO INCREASE PRODUCTION STOP YOU BETTER RECONSIDER YOUR WIRE OF THIS MORNING

          ALEXANDER BOTTS

     TELEGRAM
EARTHWORM CITY ILL 945A MAR 20 1920
ALEXANDER BOTTS
DELTA HOTEL
CYPRUS CITY MISS

OUR WIRE OF YESTERDAY STANDS STOP YOUR JOB AS SERVICE MECHANIC WITH THIS COMPANY IS GONE FOREVER STOP WE ARE PUTTING YOU ON PAY ROLL AS SALESMAN STOP TWO HUNDRED PER MONTH PLUS EXPENSES PLUS FIVE PER CENT COMMISSION ON ALL SALES

     FARMERS FRIEND TRACTOR COMPANY
     GILBERT HENDERSON SALES MANAGER

THE INDIRECT METHOD

ILLUSTRATED BY TONY SARG

THE INDIRECT METHOD

FARMERS' FRIEND TRACTOR COMPANY
MAKERS OF EARTHWORM TRACTORS
EARTHWORM CITY, ILLINOIS

JUNE 1, 1920.

MR. ALEXANDER BOTTS,
MULLER HOTEL,
KANSAS CITY, MISSOURI.

DEAR MR. BOTTS: We are informed that the road commissioners of Sillca County, Kansas, are considering buying a tractor for grading and general road work. We want you to go at once to Sandy Forks, the county seat, and sell these commissioners an Earthworm tractor.

Salesmen from other companies will probably be there. In addition, it is possible that you may encounter a little sales resistance, due to the fact that Mr. Joseph Ripley, a wealthy farmer of Sillca County, has been having trouble with his Earthworm tractor—due entirely to his own negligence—and has been blaming everything on the company.

However, we have every confidence in you and feel sure you will put over this deal. Advise us fully in your daily reports as to what progress you make.

Very sincerely,
GILBERT HENDERSON,
Sales Manager.

FARMERS' FRIEND TRACTOR COMPANY
SALESMAN'S DAILY REPORT

DATE: JUNE 3, 1920.
WRITTEN FROM: SANDY FORKS, KANSAS.
WRITTEN BY: ALEXANDER BOTTS, SALESMAN.

I got your letter yesterday, and it is a good thing you are putting onto this job a real high-powered salesman like me, rather than one of your ordinary men. When I explain the situation here, you will see that any ordinary man would have quit cold. But not Alexander Botts.

I left Kansas City bright and early this morning in my new flivver roadster, and about the middle of the afternoon—when I had gotten to within about two miles of the town of Sandy Forks—I saw something that I can only describe as a sickening sight. Right beside the main road as you approach Sandy Forks is a little bluff looking out over a very pretty lake. And right on top of this bluff, in plain sight of anybody coming along the road, was a large ten-ton Earthworm tractor with a big sign on it reading:

> **I WILL SELL THIS EARTHWORM TRACTOR VERY CHEAP OWING TO THE FACT THAT IT IS ABSOLUTELY NO GOOD, NEVER WAS NO GOOD, AND THE COMPANY THAT MADE IT WON'T STAND BEHIND IT.**
> *Joseph Ripley*

Parking my car by the road I walked up and looked at the tractor. It was old, weather-beaten and rusty, and the carburetor and magneto were gone. One of the side plates was off the crank case, so that I could look in and see that the bearings were all loose and wabbly. Running my hand up into the cylinders I could feel that they were scored and pitted in a scandalous manner.

As I sadly returned to my car there came walking along the road a young man in overalls.

"Yes," he said, in answer to my questions, "Old Man Ripley has had a lot of grief with that tractor, and he sure is sore at the company that made it. He has just got himself elected to the county board of road commissioners so he can make sure that when the county buys a tractor it will be something else besides one of these Earthworms."

"Won't the other members have something to say about that?" I asked.

"Well," he said, "old Joe Ripley is the richest and most influential farmer in the county. Most of the other members have worked for him at one time or another and they'll probably want to work for him again. They aren't apt to go against what he says."

"Where does this Mr. Ripley live?" I asked.

"Straight ahead. First house on the right."

"Thank you," I said.

THE INDIRECT METHOD

After the young man had left I pondered the situation, and as I have a quick mind I soon realized that you were right when you said in your letter that Mr. Ripley has been having trouble with his tractor. Apparently you hadn't heard that he was the main guy on the board of road commissioners, or that he was putting on this pretty little tractor show out beside the state road. But I decided you were right when you said I might possibly encounter a little sales resistance. And that is why it is lucky you sent the kind of a man that it takes to overcome sales resistance.

I didn't waste any time. I decided I would take this bull by the horns; I would beard this lion in his den. Accordingly I drove on up the road to the first house on the right, which turned out to be a nice looking white farmhouse with trees all around it and big red barns behind. In the front yard stood an old gentleman who appeared to be gazing down a well. In his hands he held a large mirror.

"Howdy, neighbor!" I said cordially, as I drew up beside the road. "What seems to be the trouble?"

"There seems to be something the matter with this well," he said, "but I can't find out what it is. It's so dark down there I can't see a thing."

"Well," I said, "it just happens that I am an expert on wells, so perhaps I can help you."

Note: As a matter of fact, all I know about a well is that it is a hole in the ground with water at the bottom. I have never even been able to figure out just how it gets there. But I felt it would be wiser to approach this guy by the indirect method and to introduce myself as a well expert rather than as a salesman for Earthworm Tractors. "My name," I said, "is Alexander Botts, Expert on Wells."

"My name," said the old gentleman," is Joseph Ripley."

"Pleased to meet you," I said. "And now we will see what is the matter with this well. First of all, what is the idea of that mirror?"

"Everybody knows," said Mr. Ripley, "that the way to look down a well is to hold a mirror at the top and shine the sunlight down into it. But this tree, right over the well here, makes such a dense shade that there isn't enough sunlight to do any good."

"Yes," I said, "that tree complicates matters, I will have to think it over and see what we can do."

At first I was going to suggest cutting down the tree, but I doubt if the old gentleman would have approved. Then, all at once, like an inspiration, I got one of the most brilliant ideas that has come into my mind for a long time.

"Have you another mirror in the house?" I asked.

"Yes," he answered.

"Fine," I said. "Go in and get it."

Somewhat doubtfully he went into the house and came out with another large mirror. I then had him go out by the road in the bright sunshine and reflect a beam of light over to where I was standing at the top of the well. With the other mirror I reflected this beam of light down into the depths. The results were extraordinarily gratifying. The inside of the well was brilliantly illuminated, and far below, floating in the water, I could see, just as plain as day, a large and probably once very handsome cat.

"Beautiful!" I said. "Splendid! I have discovered what is the matter with your well."

I then walked out into the sunlight with my mirror and had Mr. Ripley look down into the well with his. He agreed with me at once that we had discovered what was the matter with the well, and declared that the unlucky animal was no doubt a cat by the name of Cicero, which he kept at the barn and which had disappeared some two weeks before. He said he would call a couple of men from the fields to get busy and clean the well.

"No," I said, "I am an expert on wells, and I will go down for you and take out this unfortunate cat."

I will admit that I did not particularly enjoy the idea of going down into a well after a cat, but it seemed to me that this was an opportunity to get in strong with the old guy.

I immediately sat the bucket down into the bottom of the well and, after removing my coat, I went down the rope hand over hand. As I am fairly agile and as the force of gravity was working in my favor, it did not take me long to descend to the level of the water. Standing with my feet on projecting stones at opposite sides of the well, I reached down, placed all that was left of poor old Cicero in the bucket and yelled to Mr. Ripley to pull it up.

This he did. And as Cicero was mounting upwards there suddenly came to me, just like an inspiration, a plan by which I could extend my visit with Mr. Ripley and have a better chance to get in strong with him. Acting upon this inspiration I at once climbed down into the water of the well, bracing my feet against the projecting stones at the sides, until I was up to my neck in the icy water. Many men would have shrunk back from the terrible coldness of that water, but not Alexander Botts. As soon as I was completely covered by the water I began splashing and yelling as loud as I could that I had fallen in and was drowning. Mr. Ripley at once

THE INDIRECT METHOD

let down the bucket, and I proceeded to scramble up the rope, assisting myself by stepping on the projecting stones on the sides of the well.

When I reached the top I tumbled out on the grass and lay on my back, gasping, coughing, choking and rolling my eyes. Imitating as nearly as I could a person who is half drowned. Mr. Ripley set up a great hollering, and a couple of farm hands came running from the barn. He had them carry me into the house, and after they had lain me on the bed in the spare room he gave me a drink of some liquid which produced a pleasant feeling of warmth over my entire body. By the time I had had two or three drinks of this excellent stuff I was feeling very comfortable, and I was able to accept very gracefully Mr. Ripley's invitation to spend the night. He saw that my car was put in the barn, and I remained in bed for the rest of the afternoon while old Mrs. Ripley dried and pressed my clothes. Meanwhile Mr. Ripley had his men pump all the water out of the well and give it a thorough cleaning.

So far I had not mentioned tractors, but after supper I brought up the subject in a casual, and indirect way—without spilling the news that I was a salesman for Earthworms. Mr. Ripley was very willing to talk. He said that he bought the tractor last year and that after he had used it only two weeks it quit on him. He had then asked for a service man from the

When I reached the top I tumbled out on the grass and lay on my back, imitating as nearly as I could a person who is half drowned.

He gave me a drink of some liquid which produced a pleasant feeling of warmth over my entire body.

factory, and the service man had told him it was all his own fault for not putting oil in the machine, and the company had refused to repair it for him free. That made him so mad he had run it out by the road and put that sign on it so as to knock the company as much as he could.

"Did you really run it for two weeks without oil?" I asked.

"I can't remember for sure," said Mr. Ripley, "very possibly I did. But that doesn't let the tractor company out. That machine was guaranteed for a year, and it went bad after only two weeks. So it was their business to fix it up for me free. What if I did forget about the oil? Anybody is liable to forget a little thing like that. I know that I gave it plenty of gasoline and plenty of water, and anybody is liable to forget a little minor thing like oil. No, sir, these Earthworm Tractor people are nothing but a bunch of crooks."

"Well," I said, "you certainly have had a most unfortunate experience."

"Right you are," said Mr. Ripley, "and the worst of it is all the people around here are kidding me about it. I suppose I got bit once, but I ain't going to get bit again. I am a member of the board of county commissioners and we're going to buy a tractor for road work, but you can bet your bottom dollar it won't be one of those damn Earthworms. Salesmen for several other makes of tractors are going to do some demonstrating for us down around town, and we'll take whichever one of them shows up the best. There was a man here trying to sell us an Earthworm, and we told him we wouldn't even consider it."

"I didn't know there were any Earthworm salesman around here," I said.

"This man," said Mr. Ripley, "is a contractor by the name of Casey. He's just finished up a dirt moving job over in the next county at Johnsonville. He's moving out to Oregon for his next job and he has a second-hand Earthworm tractor that he is willing to sell cheap. But I told him I wouldn't take it as a gift, so that's all there is to that."

"And what are you going to do with that old tractor of yours," I asked, "just leave it out there beside the road?"

"Sure," he said, "unless somebody comes along and wants to buy it. But I haven't had any offers for it yet. I paid $6000 for it, and I'll let it go for $1000 just as it stands, including the carburetor and magneto, which I've got up here in the barn."

"Sold!" I said.

"You mean you want to buy it yourself?" he said.

"Exactly so."

"But I don't want to stick you."

"I'll risk that."

"It's absolutely no good," he said. "I have had several mechanics from town come out and look at it, and they tell me it is completely shot to pieces."

"Don't worry about me," I said. "In addition to being an expert on wells, I am also an expert on tractors, and I figure I can fix it up and maybe sell it at a profit. You said you would sell it for $1000. You certainly aren't going to be a cheap sport and back down on your promise, are you?"

"No."

"Fine," I said. "The deal is closed."

I at once got out my check book and wrote him a check for $1000 on the First National Bank of Earthworm City.

"I am dating this check June tenth—a week ahead," I said, "to give me time to transfer that much money into my checking account, but I won't take the tractor away until the check has been cashed. That will be all right with you, won't it?"

"Absolutely," said Mr. Ripley.

Shortly after this I said good night and came up to my room, where I have been spending the rest of the evening writing up this report.

From what I have related you can see that you made no mistake in sending me to handle this very difficult job. Instead of blatting out the news that I was an Earthworm salesman, and thus getting myself thrown off Old Man Ripley's farm, I have proceeded by the indirect method, and—thanks to my energy in taking advantage of the fortunate incident of Cicero in the well—I have already gotten in very strong with this old gentleman, who is the main guy on the board of road commissioners. Furthermore, by purchasing this old tractor I have taken the first step toward carrying out a very deep plan which I have evolved for the purpose of carrying through this matter in my usual brilliant manner.

    I will now close and get myself some sleep, but tomorrow I am going to start some real action around this neck of the woods.

Cordially yours,
ALEXANDER BOTTS,
Earthworm Salesman.

P.S.: In looking over the stubs in my check book, I find that my balance in the First National Bank of Earthworm City is $21.30. As I have no funds of my own to bring up this account, I must ask you to have the cashier of the tractor company place $1000 to my credit in this bank at once. Otherwise the highly important operations which I am conducting in this region are liable to be somewhat hampered.

A. B.

FARMERS' FRIEND TRACTOR COMPANY
SALESMAN'S DAILY REPORT

DATE: JUNE 5, 1920.
WRITTEN FROM: SANDY FORKS, KANSAS.
WRITTEN BY: ALEXANDER BOTTS, SALESMAN.

I was so busy all day yesterday, last night and today that I haven't had time to write you any report until just now. But when you read what I have been doing, you will realize that I have been right up on my toes all the time. I have got things moving along something swell.

    Bright and early yesterday morning I got my car out and told Mr. Ripley I was going to get some tools to repair the tractor, and that I would be back the next day. He gave me the magneto and carburetor, which I placed in the car, and then I started toward town. As soon as I had rounded the first curve I chucked the carburetor and magneto into some thick bramble bushes beside the road. Then, instead of getting any tools, I drove twenty miles over into the next county, and just outside the town of Johnsonville I found the camp of this contractor, Casey. Most of the equipment of the camp seemed to be packed up ready to move. The only man there was a young guy with red hair who told me that Mr. Casey had gone to town and would be back in about half an hour. So I had to wait, but while waiting I did not let any grass grow under my feet. I talked most pleasantly with this young redheaded guy, and I got him to show me the tractor, which turned out to be exactly the same model as the one owned by Mr. Ripley. I was also pleased to see that it had been out in the weather long enough to make it just about as rusty looking as Mr. Ripley's old wreck. The redheaded guy told me, however, that it was in A1 condition, and he proved it by cranking it up and driving around a bit for me. Then I thanked him and gave him a few cigars and patted him on the back, and talked to him so pleasantly that he finally blurted out the good news that Mr. Casey was so anxious to sell the tractor he would let it go for two thousand dollars, although he was asking and hoping for four thousand dollars.

Pretty soon after that an automobile came driving up and out stepped a rather nervous looking guy, who proved to be Mr. Casey himself. I don't know why it is, but pretty near all these contractors are nervous looking guys—they seem to have a whole lot on their mind.

I at once offered Mr. Casey six hundred dollars for his tractor, at which he let out a loud laugh and told me he wouldn't take one cent less than four thousand dollars. So we jawed around a while. I came up a little and he went down a little, until finally I told him that if he would include a big chain that was hanging on the tractor, and if he would have his man drive the tractor over to the town of Sandy Forks for me, I would pay him two thousand dollars. As he saw that he couldn't make any better bargain, and as he had already—according to the redheaded guy—decided to let her go for that, he accepted. But when I started to write out a check on the First National Bank of Earthworm City he began to look even more nervous than before.

"How do I know," he said, "that your check will be good?"

"Oh, don't worry about that!" I said. "Anybody around Earthworm City could tell you that Alexander Botts is good for a hundred times as much money as this paltry two thousand dollars."

"That's all right," he said, "but is there anybody around here that knows you?"

"No," I said, "I am afraid there isn't. But listen," I said. "This check will go through in a few days, and in the meantime I am only taking this tractor as far as Sandy Forks. You are a reasonable man and you ought to know that I couldn't skip out of the country with a great big thing like a ten-ton tractor."

"Well," he said, "I guess maybe that's right."

"Sure it's right," I said, "but, of course, if you want to call off the sale ——"

"No," he said, "I'll take a chance."

So he took the check and made me out a bill of sale.

Note: As Mr. Casey was so suspicious about my check, I did not like to ask his permission to date it ahead. Consequently it is very important that the cashier of the Farmers' Friend Tractor Company place an additional two thousand dollars to my credit at the First National Bank of Earthworm City. Be sure and have him do this AT ONCE, as Mr. Casey will probably send in the check right away, and it is possible that if this check came back marked No Funds it might very seriously cramp my style in my present undertakings and result in considerable embarrassment and detriment to myself and to the best interests of the Farmers' Friend Tractor Company.

The young redheaded guy filled up the tractor with gas and oil, and we started out for Sandy Forks with me leading the way in my car and him following with the tractor. I had him drive very slow, and we stopped a long time for lunch at a little village, so we didn't get to Sandy Forks until dark. I led the way along a road that went around the town so as not to attract too much attention, and I had him drive the tractor into a field about a mile from Mr. Ripley's farm. Then I drove him back to Casey's camp at Johnsonville in my car, after which I returned to Sandy Forks.

I had supper at a little restaurant, and about midnight I drove out to where I had left Casey's tractor. I cranked it up and then I drove it along the road until I came to Mr. Ripley's old tractor. It didn't take me more than five minutes to remove the big sign, hook onto the tractor with the big chain, and drag it down to the shore of the little lake. Then I unhooked, drove around behind it with the Casey tractor and gave it a good healthy push that sent it over the steep bank and into the deep muddy water beneath. Next, I drove back and stopped Mr. Casey's tractor exactly where the other one had been, and I put Mr. Ripley's sign on top of it. After this I walked back to where I had left my car, drove to town, got myself a room at the hotel and went to bed about two A.M.

THE INDIRECT METHOD

Bright and early this morning I drove out to Mr. Ripley's farm and found him inspecting his well, which, after being cleaned out, was now once more full of fine fresh water. "Good morning, Mr. Botts," he said as I drove up. "What is the news with you?"

"Very good news," I said. "I have just been down the country a ways to my brother-in-law's farm, and my brother-in-law has promised me that if this tractor runs as well as I expect to make it, he will buy it from me for five thousand dollars."

Note: I will admit that this statement was somewhat exaggerated. As a matter of fact, I have no brother-in-law, and even if I had I doubt very much if he would have five thousand dollars to spend on a tractor. However, I felt that the delicacy of my maneuvers in the matter of all these various tractors justified the use of a certain amount of strategy.

"If you can get five thousand for that bunch of junk," said Mr. Ripley, "you're welcome to it."

"That's fine," I said. "I was sure you'd be a good sport about it. You know that when I bought that tractor I took an awful chance, and I was ready to stand the loss in case it turned out worse than I thought it was. But now that it turns out to be in really swell condition, I figure that I am entitled to whatever profit I can make on it."

"Sounds fair enough to me," said Mr. Ripley. "But I think you are fooling yourself. The best automobile mechanics in town have told me that the machine is a wreck."

"The best automobile mechanics," I answered, "are sometimes none too good on tractors. But I am a tractor expert. I stopped off this morning and put on the carburetor and magneto, and filled the old baby up with gas, oil and water. And I find that lack of oil was really the only thing the matter. Now that I have filled her up with plenty of good fresh oil, she's practically as good as new."

"Important," said the old guy, "if true."

"If you have any plowing or any other work you want done," I suggested, "we'll crank up the tractor and try it out."

"All right," said Mr. Ripley. "I want to plow that forty-acre patch across the road. There is an eight-bottom gang plow in the barn, and if you can get that tractor up here, you can hook on and start in. But I doubt if you can make the tractor run three feet. The only way I could get it to where it is now was by hooking onto it with all the horses on the place."

"Let's go," I said, "and see what we can do."

We walked down to the tractor, and I removed the handsome sign. Then

I gave the crank one flip and the motor started with a roar. I climbed in and drove up to the barn as fast as I could, with Mr. Ripley trotting along behind—the most surprised looking old geezer I have ever seen in my life. By the time he reached the barn I had already hooked up to the plow, taken it across the road, and started a back furrow down the middle of his forty-acre field. At the end of the first round I stopped and cut off the motor, and I was very much pleased to note that I was making quite an impression on old Joe. He came up and stood behind the machine, and for a while he couldn't say a word nor do anything except open and shut his mouth in a foolish sort of way.

"I never seen the like." he said, after a while. "I wouldn't have believed it possible. Those mechanics all told me the machine was a wreck."

"These mechanics must have been trying to fool you," I said. "The idea of telling a smart, intelligent man like you that you were a dumb tractor operator! From the looks of this machine I would say that you were one of the best tractor operators in the country. You have kept it in fine shape."

"Do you really think so?" he asked.

"Sure I do," I said. "Stick around and wait until we make a few more rounds."

Mr. Ripley looked at his watch. "I wish I could," he said, "but I just happened to think I have to get to town to see a tractor demonstration that these other guys are putting on today. There are three different machines down there and they are all going to demonstrate what they can do on road grading and moving dirt."

"Where is this demonstration going to be?" I said.

"Oh, just up and down some of the main roads," he said.

"Listen," I said. "I don't want to butt in on anybody else's business, but I am an expert on all kinds of tractors, and I can give you a tip on how to find out which one of these machines is the best."

"How is that?" he asked.

"Any of these machines," I said, "can pull a grader on a nice dry road, but when you get a tractor for road work, you want one that will go through all the deep mud holes that come in the bad weather in the spring and fall. So if I were you I would pick out a nice wet swamp and make them all go through that."

"Sounds like a good idea," said Mr. Ripley. "There's a nice soft swamp just north of town."

"Make 'em go through that," I said. "And in the meantime I'll see if I can't get a little plowing done for you."

I cranked up the machine and started across the field again, and looking over my shoulder I saw Mr. Ripley climb into his car and drive off toward town.

All the rest of the morning I kept that tractor running wide open, and I certainly have got to hand it to Casey and his redheaded operator for keeping the old baby in fine shape. She went sailing back and forth across that field as smooth and steady as a ferryboat. The ground was loose and sandy and turned over just as easy as could be. I took a half an hour off at noon to eat some lunch that Mrs. Ripley gave me, and then went back and plowed all the afternoon. And at half past five I finished up the last headland and dragged the plow back to the barn.

At six o'clock, when Mr. Ripley got back from town, I was all washed up and sitting comfortably in a rocking chair on the porch. When he saw how I had plowed forty acres in one day with a machine that he supposed was nothing but a bunch of junk, I thought the poor old gentleman was going to faint. It also seemed to me that he looked just a little bit sore, so I started in to talk right away. And I will admit that I am a good talker.

"It certainly is lucky, Mr. Ripley," I said, "that you are such an intelligent, fair-minded man. Some people would be low-down enough to be mad because I am going to make such a nice profit on this tractor deal. But I know that you are a gentleman and a real good sport."

"Yes," he said, "of course I want to be a gentleman and a good sport. But at the same time I almost wish I hadn't sold you that machine. I'd almost be willing to buy it back from you for twice as much as you paid for it."

"I wish I could let you have it, Mr. Ripley," I said, "but I really don't see how I can without going back on my promise to my brother-in-law. You see I told him he could have it for five thousand dollars, and, of course, you are too fine a man, Mr. Ripley, to want me to break my word to my brother-in-law."

"I suppose you're right," said Mr. Ripley, shaking his head very sad, "but this business has gotten me all confused. I don't know whether I am going in or coming out."

"There is one thing to be thankful for anyway," I said. "You know now that you were a good judge of machinery when you bought this machine, and you also know that you are a perfectly competent tractor operator. And by the way, how did the big tractor demonstration come off down in town?"

"The tractor demonstration in town," said Mr. Ripley, "was a joke."

"Did they go through the swamp?"

"They went into the swamp," said Mr. Ripley. "At first none of them would try it, but I told them that we wouldn't buy any tractor that couldn't go through soft ground. So they all started, and all three of them are mired down so deep in the swamp that it looks like it will take a week to get them out."

"But which one of them are you going to buy?" I asked.

"I don't know," said Mr. Ripley. "We have a meeting of the board of county commissioners tomorrow afternoon to decide what to do, but I don't know as I want to buy any of those machines."

"I'll tell you how you could have a lot of fun," I said, "and show up a lot of those town people for the boobs that they are. You know that they have been saying around town that you are a bum tractor operator. Now is your chance to show them all that you are the best there is."

"How can I do that?"

"In the morning," I said, "we will both get in that old Earthworm tractor and you will drive. We will go down there and drive right through that old swamp and pull these three machines out of the mud. I guess that will show them what a real operator like you will do when you are driving a real machine, like this Earthworm."

"Do you honestly think we could do it?"

"Sure we can," I said. "These three machines are nothing but ordinary tractors, and of course they sink right down in the mud just the same as an automobile would. But this Earthworm has tracks on it like a wartime tank, and it can stay right up on top of the mud as nice as you please. And I would be glad to let you use the tractor, because I promised you I wouldn't take it away until that check came through."

"It sounds like a good idea," said old Mr. Ripley. "I believe we will do it."

"Fine," I said.

Soon after that we had supper, and as Mr. Ripley had invited me to spend the night, I came up to my room and I have been writing this report ever since.

From what I have told you, you can see that everything is going swell, and that you made no mistake in sending me to handle this very delicate situation. By the use of the indirect method I have now gotten Mr. Ripley eating out of my hand. Tomorrow morning I intend to sell him back his own old tractor—which, as I have explained, is not really his own old tractor at all. And tomorrow afternoon I will go before the road commissioners, tell them that I have just taken over the Earthworm agency, and I will then count on Mr. Ripley's help to sell them a machine for the county.

Please let me know right away whether or not you have deposited that three thousand dollars to my account in the First National Bank of Earthworm City.

Cordially yours,
Alexander Botts.

Farmers' Friend Tractor Company
Salesman's Daily Report

Date: June 6, 1920.
Written from: County Jail, Johnsonville, Kansas.
Written by: Alexander Botts, Salesman.

As you may guess from the heading of this letter, my operations in this region are not proceeding in as felicitous a manner I had hoped they would, but when I explain matters, you will see that it is not my fault. I will admit that I was very much shocked and disappointed when I visited the post office this noon and received such a chilly letter from Mr. Gilbert Henderson, Sales Manager of the Earthworm Tractor Company. I notice that Mr. Henderson says that the salesmen of this company are expected to sell tractors and not to buy them, and that the company cannot finance unauthorized purchases of secondhand tractors. Also I see that he has turned down my request that three thousand dollars be deposited to my account at the First National Bank of Earthworm City. I suppose Mr. Henderson thinks he knows how to run a sales department, but I nevertheless wish to point out that his action in this matter has probably cost the company the sale of a tractor, and has also put this particular salesman in a somewhat embarrassing position—as you may judge from the heading of this report. This is particularly unfortunate in view of the fact that I had already managed—by employing all my energy and intellect and sales experience—to bring matters almost to the point of a brilliant and highly successful conclusion.

When I think of the splendid things I accomplished this morning, it almost makes me weep to think of the depths to which I have sunk this evening. Immediately after breakfast I cranked up the good old tractor, told Mr. Ripley once more what a splendid operator he was, and had him take his place on the driver's seat. I than climbed in beside him, and we started for town. Mr. Ripley—although probably one of the worst mechanics in Kansas—is nevertheless perfectly capable of going through the simple procedure of starting and stopping a tractor, and also steering it to the right, left, or straight ahead, as the case may be. We rolled down the road very nicely and before long we had reached the swamp where the three tractors were mired down.

The salesmen and mechanics in charge of these tractors were all out with shovels and timbers trying to get them out, and there were also a great many of the townspeople, who had come to observe the excitement. I have never seen a prouder man than old Mr. Ripley as he drove that splendid Earthworm tractor out over the soft swamp in front of all the admiring townspeople. I coached him up on just what to do, and we put on a beautiful show. The salesmen for these other tractors were none too pleased to have us come out there, but they could not afford to refuse the assistance that we so kindly offered them.

First of all we drove out to the nearest tractor—a big hulk of a machine. I told Mr. Ripley to take our machine up to the front of it, and then I hooked on the big chain and told him to go ahead. It was a hard pull, but we finally got it out of its hole and dragged it up onto the firm ground

It was a hard pull, but we finally got it out of its hole and dragged it up onto the firm ground at the edge of the swamp.

at the edge of the swamp. Then we went after the other two. And just as I had predicted, the old Earthworm stayed right up on top of the soft ground and performed in a really splendid manner. After about an hour's work we had all three of these clumsy machines out of the swamp and up on the high ground.

By this time it looked as if practically all of the town was out to see the show, and when we had at last finished they all began waving their hats around and shouting with the greatest enthusiasm. Old Mr. Ripley was tickled absolutely pink. He stood up and bowed gracefully to the crowd, and when he finally sat down I told him once more that he was one of the best tractor operators in the entire United States, and also a very good sport.

"Yes, sir," he said to me, "I guess I have shown you and all the rest of them that I am a pretty swell operator, and now I am going to show you that I am also a very good sport. I will buy this tractor back from you and I will pay you exactly what you said you could get from that other guy down in the country. I will pay you the full five thousand dollars."

"No, Mr. Ripley," I said, "you are such a good friend of mine that I could not think of asking you that much. I am making a profit on this tractor myself, and I am such a good sport that I am going to let you also make a profit of one thousand dollars on this transaction. I will make it up to my brother-in-law in some other way, and I will sell you this tractor for only four thousand dollars. I know that you are offering me five thousand

dollars, but you are such a good friend of mine that I am willing to let you take it for only four thousand dollars."

It was really touching to see the gratitude with which old Mr. Ripley received my very generous offer. He shook me by the hand, and there actually seemed to be tears is his eyes as he accepted the proposition. I persuaded the old gentleman to get down out of the tractor at once, and I took him over to the bank in town while the cheers of the crowd were still ringing in his ears. With a smile on his face he wrote me out a check for three thousand dollars and handed me over my own check for one thousand dollars which he had been carrying around in his pocket. I than handed him back the bill of sale on the tractor, which he had given me, and we shook hands once more in the most cordial spirit imaginable.

Note: I wish to point out—in view of my somewhat unfortunate situation this evening, as indicated in the heading of this report—that my financial operations in respect to this tractor were rather good, if I do say so myself. You will have to admit that it takes a guy with a quick mind—and a good talker too—to buy a man's tractor for one thousand dollars and then sell back to him what he thinks is the same machine for four thousand dollars, and make him think that you are doing him a favor. And in a way I was doing him a favor—I could have soaked him for five thousand easy.

As soon as our little transaction was completed, Mr. Ripley started back to look after his tractor, and I at once turned his check into the bank, opening an account in my own name for the three thousand dollars. I felt that if Mr. Ripley calmed down from the warm glow which had been induced by the applause of the crowd, and if he then changed his mind, it would be just as well to have the check cashed so he could not stop payment.

After leaving the bank I stopped in at the post office and received Mr. Henderson's most unwelcome letter, which seemed to indicate that I had no funds at the First National Bank of Earthworm City to back up the two thousand dollar check I had given Mr. Casey. As I am a very good business man, it at once occurred to me that it might be wise to get in touch with Mr. Casey to prevent the possibility of there being any misunderstanding. Consequently, I called his camp at once on the long-distance telephone. Unfortunately, Mr. Casey was out, and the redheaded tractor operator with whom I talked did not know when he would return. As I am very conservative, I decided to take no more chances than I could help. So I told the redheaded operator to inform

Mr. Casey when he came back that I was coming over to Johnsonville at once to see him. I then went back to the bank and had them give me a certified check for two thousand dollars, after which I hired a car to take me out to Mr. Ripley's place, where I got into my own machine and drove at once to Johnsonville.

As I am always the soul of honor in all business transactions, I was resolved to avoid even the appearance of evil in my dealings with Mr. Casey. If he had not yet sent in the check on the First National Bank of Earthworm City I would redeem it with the gilt-edged certified check on the Sandy Forks bank. In case he had already sent in the check, I was prepared to place the certified check in the care of the cashier of the Johnsonville bank, to be paid over to Mr. Casey when my other check should come back.

But unfortunately, all my good intentions went for nothing. When I arrived at Mr. Casey's camp, I was met not by Mr. Casey but by a tall gentleman with a very disagreeable face. As soon as I had told him my name he pulled back the flap of his coat and showed me a cheap looking, nickel plated star, which was pinned to his vest, and which bore the words Deputy Sheriff. He then drew a large paper from the side pocket of his coat and informed me that it was a warrant for my arrest.

"I would advise you to come quietly," he said, "and I must warn you that anything you say may be used against you."

"But why," I said, "would anyone want to arrest a law abiding citizen like me?"

"This warrant," he said, "was sworn out by Mr. Casey, who charges you with passing a bad check for two thousand dollars."

"It is all a mistake," I said. "Take me to Mr. Casey and I will straighten everything out."

"Mr. Casey," he said, "left about a half an hour ago for Kansas City to see his lawyer. He said he would be back in the morning, at which time you will have a hearing before the judge."

"But I have to be back in Sandy Forks this afternoon for a very important meeting," I said, "I can't stay until tomorrow."

"Oh, yes, you can," he said.

I then gave that man all the arguments I could think of. I told him it was all a mistake. I showed him the certified check, I threatened to sue him for false arrest. I even gave him two very good cigars—which he took—and asked him as one gentleman to another to let me go for the afternoon on my word of honor to come back the next day. I talked with that man for a good half hour—and I am a pretty good talker, if I do say so myself—but it seems as though these guys with the stars on their vests are pretty hard birds to talk to. I didn't seem to get anywhere at all, and finally he even had the nerve to make me drive the both of us back to town in my own car, after which he locked me up in the Johnsonville jail.

In all fairness I must admit that this is a very handsome jail. It seems to be brand-new, and the bars of my little cage seem to be of as good quality steel as anything used even in such a high-grade machine as the Earthworm tractor. I have been scratching away on one of the bars for half an hour with a piece of a backsaw blade which I happened to have in my pocket and which they missed when they took my money and everything else away from me, but I haven't been able to make any impression on it.

So I have been sitting around in this dump all the afternoon, and now, after supper, I am writing this report with pencil and paper which the jailer's wife was kind enough to let me have. I am fairly comfortable in here, but it just makes me sick to think that I got the chief county commissioner over at Sandy Forks all worked up ready to buy an Earthworm tractor, and then I was unable to attend the meeting and put over the deal. As no Earthworm salesman was present at that meeting this afternoon, I suppose they have probably bought one of those other tractors.

Yours,
ALEXANDER BOTTS,
Sales Manager.

FARMERS' FRIEND TRACTOR COMPANY
SALESMAN'S DAILY REPORT

DATE: JUNE 7, 1920.
WRITTEN FROM: SANDY FORKS, KANSAS.
WRITTEN BY: ALEXANDER BOTTS, SALESMAN.

Well, I am out of jail again.

When Mr. Casey showed up at the hearing this morning I learned that right after he had taken my check some days ago, he had become even more nervous than usual. And as he seems to have a peculiarly low, suspicious type of mind, he had pulled a very dirty trick indeed. Instead of sending my check through in the regular way, he had had his own bank telegraph at once to the First National Bank of Earthworm City to inquire whether my account was good for two thousand dollars. When the word came back that it was not, he had at once sworn out a warrant for me. However, when I explained that it was all a mistake and produced the certified check, he at least had the decency to say he would drop the charges if he was sure to get his money. So as soon as they had telephoned to the bank in Sandy Forks and found that the certified check was OK they turned me loose.

I drove back to Mr. Ripley's house outside Sandy Forks with a heavy heart. I was somewhat reassured when that gentleman met me with a smiling face, and I was greatly pleased when I heard what had happened at the meeting.

It seems that the commissioners had unanimously turned down the three other tractors. And, as there was no Earthworm salesman present, they decided—on Mr. Ripley's motion—to write in to the Farmers' Friend Tractor Company and order two ten-ton Earthworm tractors to be used in county road work. Their letter will, no doubt, be in your hands by the time this report reaches you. I am leaving tonight for Kansas City, and I wish to point out that although I have used the indirect method in this transaction so that my name does not appear on the county order, and although nobody around here suspects that I am a salesman for the Earthworm Company, nevertheless I am entitled to my regular commission on this sale.

As I look back on the events of the last three days I am very much impressed with the energy and resourcefulness I employed in bringing this transaction to such a successful conclusion.

> Very cordially yours,
> ALEXANDER BOTTS,
> *Salesman.*

P.S.: I have been wondering what I ought to do about that extra one thousand dollars in the bank that seems to be left over. If the Earthworm Company had put three thousand dollars into this financial operation I would feel that this one thousand dollar profit belonged to the company. But as long as the company did not do so, I have decided, after long thought, to keep the money myself.

I'M A HARD-BOILED BOZO

ILLUSTRATED BY TONY SARG

I'M A HARD-BOILED BOZO

FARMERS' FRIEND TRACTOR COMPANY
MAKERS OF EARTHWORM TRACTORS
EARTHWORM CITY, ILLINOIS

JANUARY 3, 1921.

MR. ALEXANDER BOTTS,
WHITESTONE HOTEL,
CHICAGO, ILLINOIS.

DEAR MR. BOTTS: At your earliest convenience we would like you to go to Centerville, Wisconsin, and call on Mr. Edward Beekman, a farmer, who lives near that place. Last spring Mr. Beekman bought one of our ten-ton Earthworm tractors at six thousand dollars, paying three thousand cash and giving us his notes for three thousand balance, secured by a chattel mortgage on the tractor.

The first of these notes for one thousand dollars came due last September, and in response to our letters requesting payment we have received evasive replies, but no cash.

You will call on Mr. Beekman and demand immediate payment, failing which you are hereby authorized to take possession of the tractor, which you will have shipped back to the factory for overhauling and resale, as per our agreement.

Although we do not usually ask our salesmen to undertake collections, we are asking you to do it in this case because we have no one else available at the moment, and we feel sure you will be willing to undertake this job and will carry it through with your usual energy.

We enclose Mr. Beekman's note and the mortgage. Be sure to give us full daily reports of what you do.

Very truly,
GILBERT HENDERSON,
Sales Manager.

Farmers' Friend Tractor Company
Salesman's Daily Report

Date: January 5, 1921.
Written from: Centerville, Wisconsin.
Written by: Alexander Botts, Salesman.

I received your letter in Chicago yesterday. I have come right up here to Centerville, and I wish to state that I am perfectly willing to tackle this little job of collecting. What it takes to get money out of these low-down dead beats, I've got. In general I try to cultivate a polite and ingratiating manner, but on a job such as this I'm a hard-boiled bozo.

I will go after this Beekman guy like a mad bull, and before I leave I will either have the money or take the tractor away from him.

And it certainly is lucky you sent a guy like me who is not afraid of difficulties, because when I explain the situation up here, you will see that any ordinary man would have quit cold.

I suppose when you sit in your swell steam heated office in Earthworm City and dictate a letter telling me to call on a bird who lives just outside of Centerville, Wisconsin, you think you are asking for something very simple and easy. But you don't know what this country is up here. It is practically a suburb of the North Pole, and yesterday and last night they got one of the biggest blizzards in the memory of man.

When I got off the train this morning—four hours late—I found the town practically covered up, except for one little path that had been shoveled from the station across the street to the hotel. The snow seemed to be about up to my neck on the level, and where it was drifted it looked like the pictures of the Himalaya Mountains. It is so cold I don't dare look at a thermometer to find out what it really is.

I followed the little path across to the hotel and asked the clerk if he knew where Mr. Edward Beekman lived. He replied that Mr. Beekman lived about ten miles south of town on the main road leading to a place called White Creek. But when I asked how I could get there, he said he didn't have the faintest idea.

"When the roads are open," he said, "there is a bus line that runs all the way down to Milwaukee; you can hop a bus at the hotel here and get off right in front of his house. But the buses have been tied up since November, and now you can't even telephone, as the blizzard has broken down the lines."

"Can't I hire a car," I asked, "or a horse and wagon?"

At this the clerk let out a loud laugh. "Nobody ever tries to run a car around here in the wintertime," he said. "We all put our machines in storage in the fall and don't take them out till spring. Usually the farmers break out the roads with horses after every storm. But there is so much snow this time that it will be several days before you can get out into the country."

"Several days, my eye!" I exclaimed. "I can't wait. I am going to find some way to get out there today."

And it gives me great pleasure to report that this remark was no idle boast. I have bought myself—in a store next to the hotel—a pair of high-grade snowshoes, together with all necessary accessories. Although I had never used these contrivances before, I have—in ten minutes practicing behind the hotel—discovered that I am a natural-born snowshoe artist. I have been told that Mr. Beekman's house is straight south on a direct road, and I figure that if the road is concealed by drifts, I can follow the telephone poles, which ought to be high enough to stick up through the snow. You see I think of everything. And now that I have had dinner at the hotel and practically finished this report, I am all ready and raring to go.

I have given you this very full account of the snowbound situation here so that you will cheerfully OK my expense account, which will on this occasion contain charges for my Arctic outfit, consisting of one pair snowshoes, one pair leather moccasins, one fur cap with earlaps, one pair leggings, one pair fur mittens, one Mackinaw coat, one pair lumberman's socks, and one suit extra heavy, red flannel underwear. By beating down the storekeeper, I was able to get the whole business for the very moderate price of $93.49.

Cordially yours,
ALEXANDER BOTTS,
Salesman.

Farmers' Friend Tractor Company
Salesman's Daily Report

Date: January 5, 1921, 9 p.m.
Written from: The Beekman Farm, Outside
   Centerville, Wisconsin.
Written by: Alexander Botts.

Well, I followed the telephone poles and I got here. I will not describe my trip in any way except to state that my splendid new clothes kept me comfortably warm, and that the snowshoes are not so good after all. Even for a born expert like myself, it is impossible to walk in anything like a natural manner with a couple of tennis rackets strapped to the feet. The only thing a guy can do is to waddle along like a duck, and a pretty fat old duck at that. A very fatiguing procedure.

After I had waddled the first mile I was very tired. After I had waddled the first five miles, I will admit that I was practically all in. But I never thought of turning back. Calling upon every ounce of my will power, I kept bravely on, and just at evening I had covered the entire ten miles and found myself in front of a very pretty little white farmhouse, set in among the snow covered hills. Behind the house was a large red barn, and in front, sticking up through the snow, was the top of a red gasoline filling station pump.

I tottered up to the door of the house, knocked somewhat feebly, and was admitted by a young lady. Even in my exhausted condition, I could not help noticing that she was what I would call very good-looking. She was of medium size and slender, but with a gracefully athletic build. She had blue eyes and beautiful golden hair of about the same color as the very highest grade of light cylinder oil.

As I am a hard-boiled bozo, however, I scarcely noticed these points, but came directly to the business in hand.

"I wish to see Mr. Edward Beekman," I said.

"Come right in," she replied, smiling very cordially, "and sit down by the fire. You must be terribly cold. How far have you come?"

"I have come from Centerville," I said

"You poor thing!" she went on. "What a frightfully long trip. You must be pretty strong and husky, or you never would have made it."

"Yes," I admitted, "that is true." But as flattery such as this has no effect on me, I at once brought the conversation back to the matter in hand. "I wish to see Mr. Edward Beekman," I repeated.

The only thing a guy can do is to waddle along like a duck.

The young lady called upstairs to someone whom she addressed as "Ted," and there shortly appeared a man who was large and powerful, but very young looking—nothing but a kid.

"I want to see the man who owns the Earthworm tractor," I said.

"I guess I'm the one," he replied.

"All right," I said. "My name is Alexander Botts."

"Glad to meet you, Mr. Botts," he remarked. "My name is Edward Beekman, and this is my wife."

I then shook hands with both of them, and I will admit that it is not as easy as you might think to get hard-boiled with mere children such as these people seemed to be, especially when they were such nice looking children. However, with me business is always first.

"I represent the Farmers' Friend Tractor Company," I said, "the people who sold you your Earthworm tractor—and I have come to collect the thousand dollars which was due on this machine last September, and which has not yet been paid."

At this the two children looked very much embarrassed, and Mr. Beekman said, "I am sorry, but I just haven't got that much money."

"How much have you got?" I asked, in my practical way.

"Eight hundred dollars."

"Isn't there some way you can raise the other two hundred?"

"I'm afraid not," he replied. "I really need this eight hundred for a lot of overdue local bills; but I would be willing to let you have it, if you would trust me till next fall for the rest."

"By next fall there will be two thousand dollars more due," I reminded him.

"If you would only let me off now, I might be able to pay the whole business by that time."

"I am sorry," I said, talking as hard-boiled as I could, "but I have no authority to grant you any such extension. I have the mortgage and your note here in my pocket, and if you cannot pay I will have to take the tractor."

"What are you going to do," he asked—"drive it back to town tonight?"

"To tell the truth," I replied, "I haven't figured out yet just what I'll do."

"Better spend the night with us," said Mr. Beekman, "and in the morning you can see how the weather looks and decide what to do. I hate like the devil to lose the tractor, but I can't blame you for taking it, so I might as well be a good sport. You can have it any time you want."

After rapidly considering the situation I came to the conclusion that I did not like the idea of starting out again in all the snow, so I decided to spend the night. Mrs. Beekman soon set out a most excellent supper that proved she was a swell cook. And my long waddle through the snow had given me an appetite that permitted me to absorb a really surprising amount of food.

After supper we sat around a large open fire. As the hard-boiled business part of my visit had already been completed and as these people were treating me so much better than I could have expected, I decided to be as agreeable as possible. I entertained them with some of my very best Swedish jokes.

And afterward we talked of one thing and another and became very good friends indeed, considering the nature of my visit. By the end of the evening they were calling me "Alex," and I was calling them "Ted" and "Anne," and I had found out how it happened that they were stuck way up here in Wisconsin with a tractor they couldn't pay for.

It seems both of them are just twenty-two years old, and they were born and brought up in Chicago. Ted had been in the Army in France, and when he came back he had worked in some office where he had met Anne, who was a stenographer in the same place.

Like lots of other city people, they hated city life and wanted to live in the country and run a farm. They decided to try it.

Ted had inherited five thousand dollars from his father, and they spent practically the whole thing in making down payments on the farm and on various pieces of equipment, including the big ten-ton Earthworm. And early last spring they threw up their jobs, got married, and came up to their farm.

"How big is your place?" I asked.

"Fifty acres of cultivated fields," said Ted, "and a hundred acres of woods."

"Holy Moses!" I said. "You bought a great big ten-ton tractor to cultivate fifty acres! You are a bigger sap than I took you for, Ted. You could plow your whole farm in two days easy, and then you would have that great big expensive machine sitting around idle the rest of the year."

"I know," said Ted. "I hoped I could get work plowing for other farmers or grading the roads. But all the farmers around here do their own plowing, and the road commissioners are so old-fashioned they won't use anything but horses."

"Too bad I wasn't here; I could have talked them into giving you a job," I said. "But probably you are not much of a salesman. Did you know anything about running a tractor when you first bought your machine?"

"I sure did. I ran an artillery model Earthworm when I was in the Army. I decided then that an Earthworm was the finest tractor in the world. And maybe that is the real reason I bought one—just for the pleasure of having such a swell piece of machinery around."

"I know just how you feel," I said. "The Earthworm is indeed a wonderful machine. But your admiration isn't going to help you pay for it. Haven't you made any money out of your farming?"

"Not much. We've made enough to keep ourselves alive, but we haven't saved a cent. That eight hundred dollars I spoke of is what is left of the five thousand we started with. And we have lots of debts. Besides what we owe on the tractor, we have a five hundred dollar payment on the farm mortgage that is past due. Then there are a good many little bills, and we owe the oil company two hundred dollars."

"What's that for?"

"That was another good idea gone wrong. We thought we could make a lot of money out of all the automobile traffic that goes along this road. But we didn't get our filling station installed until late in the fall, just before the snow came and blocked the road, so we have over five hundred gallons of gasoline and a whole lot of oil on hand, with no chance of selling it until spring."

As I am rather quick on business matters, I was beginning to suspect that Ted's financial situation was not as sound as might be desired.

"It seems to me," I said at length, "that you are in rather bad shape."

"I'm afraid we are," said Ted. "You are going to take the tractor away, and I suppose they will foreclose the farm mortgage in the spring. Then Anne and I will have to go back to Chicago, and I'll probably have to work about five years in some filthy office before we can save enough to come back here and try again."

"So you aren't completely discouraged with farming?"

"I should say not," spoke up Anne. "We've been happier here than ever before in our lives. We are both crazy about the country, and we think it is the only place to bring up children. You see, we hope—that is, we think—I mean, we expect that next summer we will have—that is, there will be an addition—there will be three of us."

I have put in all the above dashes to indicate hesitation, so you can see that Anne is really a very nice, old-fashioned kind of girl, and is properly somewhat reluctant to discuss these delicate subjects—even with such a friend of the family as I have become.

After I had congratulated them on the coming event, I tactfully changed the subject by asking if I could see the tractor.

"Sure," said Ted. "Old Betsey—that's what we have christened her—is right out in the barn." He lit a lantern, and Anne and I followed him out.

I was pleased to see that Betsey was in fine condition—as bright and shiny as the day she left the factory. Ted and Anne both seemed very proud of her, and it filled me with sorrow to think that I was about to take this beautiful machine away from these two excellent people. After we

I'M A HARD-BOILED BOZO

had finished looking over the tractor we returned to the house and Anne showed me up to my room. Since then I have been writing this report, which I will mail as soon as I get a chance.

I have made a very full report of the situation here, so you can see just what sort of a problem I am up against. And I wish to point out that you did very well in tending a hard-boiled bozo like myself to handle things. If you had sent a softhearted, sentimental guy, he would have been so overcome at seeing such nice young people as Ted and Anne in such a hard fix, that he would have let them hang onto their tractor whether they paid for it or not. This would have been very wrong, as it would have made them into dead beats and would have left the company holding the sack.

On the other hand, I must admit that I rather hate to take the tractor away. I always try to look at things from all points of view, and it occurs to me that it is not good for the reputation of the Farmers' Friend Tractor Company to have it known that an Earthworm owner had gone busted because he could not find enough profitable work for his machine to do. The problem as I see it is to find some way by which the Beekmans can get hold of enough money to climb out of their financial hole and keep their tractor. At present it does not look as if there was any way to do this. But if there is a way, you may be sure that Alexander Botts will find it.

I will now go to bed, and while I am asleep I will let my subconscious mind tackle this problem. And tomorrow morning, when I am rested and refreshed from my long waddle through the snow, it is my intention to start things moving around here, and stir things up in such a way as to get some real results.

Very truly yours,
ALEXANDER BOTTS,
Salesman.

Farmers' Friend Tractor Company
Salesman's Daily Report

Date: January 6, 1921, 9 p.m.
Written from: The Beekman Farm.
Written by: Alexander Botts.

It gives me great pleasure to report that I have put in a very busy day, and have made a good start toward bringing matters to a satisfactory conclusion.

You will remember that last night the prospects looked very gloomy indeed. Here were these two excellent young people on the way to losing their tractor and their farm. Here was the entire countryside presenting the most disheartening picture; all the roads blocked with snow, all the automobiles put up for the winter, and all the merchants of Centerville so snowbound that they had practically no business at all. Everything seemed all wrong.

But when I sprang from my bed bright and early this morning, my heart was full of joy and hope. For, just like an inspiration, there had come into my mind a scheme for doing away with all this sorrow and grief.

The scheme was simple but magnificent. The chief thing that the inhabitants of this benighted land need is to have the snow plowed off their roads. The chief thing that these splendid Beekman children need is a good paying job for their tractor. The answer is self-evident.

As soon as I could get into my clothes, I rushed downstairs. "Ted," I yelled, "it is all settled! We are going into the snowplowing business."

"How do you mean?" he asked.

"We are going to put a snowplow on Old Betsey's nose, and we are going to plow the roads, and we will make all kinds of money."

"Who is going to pay us?"

"The county road commissioners."

"I doubt if they will," said Ted. "I made a proposition to them last fall that I would keep the roads cleared for them this winter. But they said it wouldn't be practical; they never had done it, and they didn't intend to do it."

"That's what they think now," I said, "But wait till they see what sort of work we can do. We will plow the main road all the way from Centerville to this town of White Creek free of charge. And when they see how good it is they will give us a regular job."

"I tried to talk them into it last fall," said Ted, "but they are too stubborn and set in their ways."

"Wait till I get after them," I said. "With a sample of our work to show them and a flow of languages such as I have got, they will soon be eating out of our hand."

"I'm not so sure about that," said Ted, "but I would just as soon try."

So, immediately after an excellent breakfast—which proved once more that Anne is a swell cook—I started things moving in my usual energetic fashion. With Ted and Anne to help me, I gathered together a lot of timbers and planks from around the barn. Then we all worked fast and furious, and by noon we had a rough, but large and imposing locomotive type snowplow rigged up on the front of Old Betsey. We filled the old lady up with gas and oil from the filling station, after which we had lunch. As I said before, Anne certainly understands the art of cooking.

After lunch Ted cranked up, and we both climbed aboard and started. The grousers on the tracks gave us splendid traction and, as Ted had taken very good care of the motor, we had all the power we needed. We nosed our way through a tremendous drift in front of the barn, swung out into the road, and opened her up wide, while Anne waved encouragement from

By noon we had a rough, but large and imposing locomotive type snowplow rigged up on the front of Old Betsey.

the front porch. The snowplow worked to perfection—great mountains of snow rolled off to the right and left—and with the motor roaring like an airplane, we moved majestically forward at three miles an hour in the direction of White Creek.

As this was strictly a moneymaking venture, I was determined to pick up as much on the side as I could. Consequently I stopped opposite the first farmhouse I came to and floundered through the drift up to the door. An old man with a beard answered my knock, and I asked him if he would like to have his driveway plowed back as far as the barn.

"The main road is going to be plowed all the way from Centerville to White Creek," I explained, "and if you have your driveway plowed also, you can take your car out and drive to town just as easy as if it was summertime."

"It's a fine idea," said the old man. "Go ahead and plow her out."

"The charges," I said, "will be ten dollars. Just slip me the cash and the work will be done in the twinkling of an eye"

"Ten dollars!" said the old man. "I should say not. I'd rather shovel it myself."

"Very well," I said. "Goodbye."

The old man shut the door, and I returned to the tractor. As I am quick at sizing up a situation, I soon came to the conclusion that ten dollars was perhaps a little too much to ask. Consequently I quoted a price of five smackers at the next farmhouse. But even this seemed to be more than they cared to pay. So at successive houses I dropped down to four, three, two, and finally one dollar. At this last price I picked up jobs from about half the people interviewed.

This stopping at each house delayed us a good deal, so it was five o'clock in the afternoon, and already getting dark, when we arrived at White Creek. Accordingly, after I had dropped my yesterday's report into a mailbox, we turned around, put the machine in high, and came clattering back along that beautifully plowed road as fast as we could—arriving at the Beekman place at about eight o'clock.

I had no use for my snowshoes on this trip, but as the weather was distinctly cool, my newly purchased Arctic garments, including the red flannels, were all that saved me from freezing to death. I mention this fact so that you may recognize my wisdom in buying these articles, and so that it will be easy for you to OK the expense account on which I have charged them.

Upon arriving at the house we were warmly welcomed by Anne.

On counting up the money we had made by plowing driveways, I found we had fifteen dollars—just about enough to pay for the gas and oil.

"But we haven't really made expenses," I said. "We ought to figure about ten dollars a day for interest and depreciation on the machine, and quite a bit more than that to pay us for our valuable time."

"Yes," said Ted. "Considering that this machine will do the work of eighteen or twenty horses, we ought to get at least fifty dollars a day. And at that price we would really be making money.

"Right you are," I said, "and I figure that the amount of plowing we can do in a day ought to be worth at least a hundred dollars a day to the inhabitants of this snow infected region. Consequently," I went on, "we will clear the road to Centerville tomorrow, and we will put it up to the county commissioners that it is their duty to pay us that much for our services. One hundred dollars a day, or, if they prefer, three dollars and a half a running mile."

"I hope you can convince them," said Ted.

"Trust me," I replied. "I am one of the best little old talkers in the whole United States. As a persuader, I am good, and I admit it myself."

With these words I came up to my room, and now that I have finished my daily report I will retire with high hopes for the morrow.

Yours enthusiastically,
ALEXANDER BOTTS,
Salesman.

FARMERS' FRIEND TRACTOR COMPANY
SALESMAN'S DAILY REPORT

DATE: JANUARY 7, 1921, 1 P.M.
WRITTEN FROM: CENTERVILLE HOTEL,
 CENTERVILLE, WISCONSIN.
WRITTEN BY: ALEXANDER BOTTS.

It gives me great pleasure to report that the day's activities are proceeding in an unusually auspicious manner. Bright and early, after a really swell breakfast, Ted and I went out and twisted Betsey's tail. As was to be expected, she started up with a beautiful roar. Waving goodbye to Anne, we swung out into the road and headed for Centerville. On the way we collected twenty dollars for plowing out driveways, arriving in town at about eleven o'clock.

On this trip I did the driving myself, and it is lucky that this was the case, as we very nearly had a serious accident which was only averted by my coolness and skill. This near accident was caused by the fact that the weather was not as cold as yesterday, so that, after we had driven about a mile, my Arctic garments began to be a little too warm. This warmth, in conjunction with the somewhat rough, woolly texture of my red flannels, produced a condition which made it absolutely necessary for me to twist about in my seat and scratch various parts of my person.

While I was reaching with my left hand for a point on my right shoulder blade, and while, as a consequence, my eyes wandered momentarily from the road, there came a sudden, sickening crash. Quick as a wink I grabbed the levers, stopped the tractor, and then backed up a few yards. A single glance of my practiced eye told me what had occurred. The machine had veered to the side and had run into the railing of a concrete bridge over which we chanced to be passing at the time. Had I been a less skillful driver, or had I waited one-tenth of a second more before stopping, we would have gone right on through the railing and dropped to the frozen surface of a stream twenty feet below. As it was I escaped with no damage at all to the tractor or the sturdy plow, and without knocking off more than about fifteen feet of the concrete railing.

When we got to Centerville we were most fortunate in finding the county road commissioners gathered at their regular monthly meeting. I at once introduced myself, explained what I had been doing, and persuaded them all to come out and look at the tractor and the results of the plowing. I then brought them back to the courthouse and offered to plow as many of the county roads as they wanted for a flat price of one hundred dollars a day or three dollars and a half a running mile.

As they seemed a little hesitant I started in and made one of the finest orations I have ever put across. I compared the paltry three dollars and a half that I would charge with the hundreds of dollars it would cost to shovel a mile of road through these drifts by hand. I told them that a plowed road would dry out so quickly in the spring and be in such good condition that the plowing would pay for itself three times over by saving most of the spring road scraping. I pointed out the tremendous loss suffered by the merchants of the town through the inability of the farmers to come in and do their trading when the roads were impassable.

And after going into these and other economic aspects of the case with great thoroughness, I concluded with a tremendous emotional appeal on sentimental and humanitarian grounds. In the choicest English at my command, and with many graceful gestures, I pictured the case of a beautiful child taken suddenly ill in a farmhouse far out on one of those snow blocked roads. I pictured the weeping mother and the desperate attempts of the father to telephone the doctor. In vain! For the telephone wires are down, and owing to the condition of the roads, it is impossible to repair them. Furthermore, even if a message could be sent to town, the doctor would not be able to make his way out to the farm through the terrific drifts that block the way. I then pictured the heroic father rushing out into the storm and the night to seek help in this great emergency. But again in vain! For the father freezes to death in a first snowdrift, while his beloved little one perishes for lack of proper medical attention.

"But if the roads had been plowed," I said, "the father could have telephoned, the doctor could have sped out in his high-powered car, and two precious lives would have been saved."

As I finished this recital I was practically in tears, and I noticed that several of the commissioners were using their handkerchiefs. The chairman said he was much impressed, and requested me to withdraw while they deliberated.

"But if the roads had been plowed," I said, "the father could have telephoned, the doctor could have sped out in his high-powered car, and two precious lives would have been saved."

I accordingly took Ted over to the Centerville Hotel, where we have had a dinner that was not so good, and where I have been writing this report. As it is now mail time, I will close, and in my next communication I expect to report that Mr. Ted Beekman has started to make so much money that we need have no further worry about the payment of his notes. And thus will come to a close another brilliant chapter in the record of my services to the Farmers' Friend Tractor Company.

Cordially yours,
ALEXANDER BOTTS,
Salesman.

FARMERS' FRIEND TRACTOR COMPANY
SALESMAN'S DAILY REPORT

DATE: JANUARY 7, 1921, 9 P.M.
WRITTEN FROM: THE BEEKMAN FARM.
WRITTEN BY: ALEXANDER BOTTS.

It gives me great pain to report that a train of distressing circumstances entirely outside my control has for the moment delayed my activities in this region. I will explain exactly what has occurred, so that you can appreciate what I am up against and see that the present regrettable situation is not in any way my fault.

The first unfortunate incident was the almost incredible action of the road commissioners. You will scarcely believe me when I tell you this, but it is, nevertheless, a fact that these men—after listening to my masterly address, which had them all practically in tears—proceeded to put away their handkerchiefs and unanimously vote against spending any of the county money on such an unheard-of activity as plowing the roads.

By the time Ted and I called at the courthouse the commissioners had already adjourned and gone home, and there was nobody there except the county clerk, who told us this shocking news. He further informed us that the commissioners had in some way been informed that we had knocked off fifteen feet of railing from the bridge outside of town. And they had directed the county clerk to hand Ted a bill for two hundred dollars to cover the cost of repairs.

I'M A HARD-BOILED BOZO

This last unkindest cut of all was almost more than I could bear. But I never faltered. I decided to save what I could from the wreck of our lost hopes.

"Let us drive Betsey back to your house." I said. "Then you can give me that eight hundred dollars before these worthless commissioners get hold of any of it, and I will see that you are allowed to keep your tractor till spring. As soon as the weather opens up you may be able to get enough work for the machine to pay the rest you owe. It is a slender hope, but all we have."

Ted at once agreed to this proposition, so we drove back to the farm, where Anne greeted as with a glad, happy smile, and reported that the filling station had been doing a rushing business. Apparently the news that we had cleared the road had spread rapidly; people had got out their cars, the motor bus had started running, and Anne had sold twenty-eight dollars' worth of gasoline and oil.

But our joy at this news was short-lived, for Anne proceeded to tell us that there had been other visitors besides the customers for gasoline. The man who owns the mortgage on the farm had had so little sense of decency that he had actually taken advantage of our snowplowing and come out in his flivver to demand a payment of five hundred dollars which was a month overdue. Furthermore, the manager of the oil company had had the bad taste to appear with a demand for two hundred dollars due him since last fall for gas and oil. And poor Anne seems to have no business sense. She had actually signed checks to pay both of these bills, so that the eight hundred dollars bank balance was now reduced to one hundred.

This was a stunning blow, but I could not very well bawl out Anne about it, as I had already refused this eight hundred dollars when I first came, on the ground that I would take a thousand or nothing. And Anne, in her innocence, had supposed that I really meant it.

Ted offered me the remaining hundred, and remarked that if I wanted it I had better take it at once, as there would probably be other bill collectors after it first thing in the morning. I said I would think it over. After carefully pondering the situation, however, I finally came to the conclusion that it would be useless for me to take this paltry hundred berries. If these people could not even make a decent start toward paying for their tractor, the most merciful course would be to end the agony as soon as possible and take the machine away from them. But as it was getting late, and as there was no way to get back to town, I resolved to take no action for the moment. I accordingly partook of an excellent supper, hiding my dreary thoughts under a jovial exterior, and cheering up poor old Ted and Anne with an entirely new bunch of Swedish wise cracks.

I have now come up to my room, where I have been writing this report. And tomorrow morning, although I hate to do it worse than anything I have ever done in my life, I will take possession of that tractor like the hard-boiled bozo that I am.

Yours,
ALEXANDER BOTTS,
Salesman.

FARMERS' FRIEND TRACTOR COMPANY
SALESMAN'S DAILY REPORT

DATE: JANUARY 8, 1921.
WRITTEN FROM: THE BEEKMAN FARM.
WRITTEN BY: ALEXANDER BOTTS.

Well, I have been hard-boiled. I have carried this matter through to a conclusion, and I am leaving for Chicago on the noon train. When I relate exactly what has occurred, you will see that I have done my duty, and done it pretty damn well, if I do say so myself.

Bright and early this morning I went downstairs all ready to announce that I was taking the tractor away. But Ted and Anne greeted me so cordially that I decided it would be better to wait until after breakfast. And by the time I had finished this meal—which was as usual most excellently cooked—I felt in such a kindly frame of mind that I decided to wait a little longer. Consequently, leaving Ted and Anne in the kitchen, I walked into the front parlor and spent about five minutes scowling at myself in a mirror so as to get worked up into a mean state of mind.

About the time I had completed this exercise and got myself all ready for some real dirty work, but before I had had a chance to go back and start in on Ted and Anne, I chanced to glance through the window. A luxurious motor car had stopped just outside, and a large and important looking man in a tremendous, expensive fur coat was coming up to the house. I promptly opened the door, and he asked to see Mr. Edward Beekman. As I was—as I have explained—in a hostile frame of mind, I lost no time in telling him exactly what I thought of him.

"You dirty bill collectors make me sick and tired," I said.

"You dirty bill collectors make me sick and tired," I said. "Why can't you leave these poor people alone? Isn't it enough that you are already lousy with wealth? Look at your elegant car! Look at your disgustingly expensive coat! Think of the money you have probably extorted from widows and orphans! Then think of these two splendid young people you are hounding, and if there is a spark of decency left within you, your fat face will be suffused with a blush of shame!"

I have repeated exactly what I said, so that you can see I am still one of the best talkers in your whole organization—although in this particular instance it appears that I did not know what I was talking about.

The man in the fur coat appeared somewhat taken aback. "Let me explain," he said. "Let me explain."

"Very well," I replied, "but make it snappy."

"My name is George Westerville," he said. "I am the president of the Central Wisconsin Autobus Company, and I wish to hire Mr. Beekman to do some snowplowing for me."

"Step inside, Mr. Westerville," I said.

He entered and took the chair I offered. Ted and Anne came in from the kitchen at this moment, but I motioned them to keep out of the discussion.

"I am Mr. Beekman's partner, Mr. Westerville," I said, "and I am the guy you talk business with."

I sat down, and he explained that he ran five hundred miles of bus lines, covering a good portion of the state, and that he suffered tremendous losses when the roads were blocked. The snow in Wisconsin, he said, was too heavy for horse-drawn plows and for plows on trucks and buses. He had never heard of the Earthworm tractor before, but since he had seen the work we had done, he was convinced it was the only thing for him. He had tried to get the state and county commissioners to buy some of the machines, but he had had no luck, so he had decided it would pay him to do the plowing himself.

He had tried to buy an Earthworm, but the dealer in Chicago had wired that he couldn't promise delivery for several weeks. He wanted action at once—he was losing money every day—and he wanted to buy Mr. Beekman's tractor or else lure him to plow out the roads.

While Mr. Westerville was explaining these things I was watching him very narrowly, and as I am a wonderful judge of men, I detected that my aggressive greeting had forced him into a somewhat apologetic frame of mind. His manner betrayed that he actually thought we would be doing him a great favor to work for him. I at once decided that it would be just as easy—now that I had started the day as a hard-boiled bozo—to keep on the same way. I recalled the words of General Hines in the Battle of the Argonne: "Now is the time to strike, and strike hard."

"I am sorry, Mr. Westerville," I said coldly, "but the tractor is not for sale. Furthermore, we have taken a contract to haul logs for the Eureka Wooden Box, Barrel, Kitchen Cabinet and Furniture Manufacturing Corporation up in the northern part of the state. If we are behindhand in this work we will forfeit a bond of one thousand dollars which we have posted with them. So that is that. Kindly close the door as you go out."

Note: Perhaps I should explain that, as far as I know, there is no such company as the Eureka Wooden Box, Barrel, Kitchen Cabinet and Furniture Manufacturing Corporation. But I thought that as long as I was evolving a name, it might as well be a good one.

For a moment I was afraid that Mr. Westerville might actually go out—closing the door as I had suggested—but such was not the case. He became even more apologetic and pleading than before. He stated that he absolutely must have this plowing done.

But as I contemplated that elegant fur coat and figured on the money it must have cost, I became more and more disagreeable. And when I finally yielded, I wrote out and made him sign one of the prettiest little contracts I have ever seen.

It provided that Ted was to hire extra operators, and run three shifts, so that he could plow day and night and get over the whole five hundred miles of bus lines in something like a week. Payment was to be made every Saturday night at the rate of six dollars a running mile—which was not so bad, in view of the fact that the day before we would have been glad to get three and a half from the county commissioners. Furthermore, the contract was to be in force two years—Ted to plow the roads after every snowfall exceeding six inches in depth, U.S. Weather Bureau figures. And in addition to all other payments, Mr. Westerville was to give us at once his check for one thousand dollars, as a bonus to cover the bond to the Eureka Wooden Box, Barrel, Kitchen Cabinet and Furniture Manufacturing Corporation.

When Mr. Westerville read this contract he let out a faint groan. But I pointed out that his increased profits would more than cover his payments to us; and I further cheered him up by telling him that out of pure generosity we wouldn't charge him anything for the thirty miles we had plowed already; and he finally signed the contract and the thousand-dollar check.

After he had gone Ted endorsed the check over to me, so he is now all paid up to date. As he will take in about three thousand dollars on this first plowing; as there will undoubtedly be other snowstorms both this year and next; and as the gasoline filling station is now doing a brisk business, he will have no trouble in paying the two thousand dollars which he still owes on Old Betsey.

It is now time to leave for the railroad station, so I will not be able to go into any greater detail regarding the energy and resourcefulness I have displayed in handling this little collection affair. In conclusion I wish to state that I am leaving these two splendid young people in the best of good spirits, and I may add that they have quietly informed me that they hope it will be a boy, so that it can bear the brave name of Alexander Botts Beekman.

> Proudly yours,
> ALEXANDER BOTTS,
> *Salesman.*

THE WONDERS OF SCIENCE

ILLUSTRATED BY TONY SARG

FARMERS' FRIEND TRACTOR COMPANY
MAKERS OF EARTHWORM TRACTORS
EARTHWORM CITY, ILLINOIS

MARCH 4, 1921.

MR. ALEXANDER BOTTS,
GIFFORD HOTEL,
OMAHA, NEBRASKA.

DEAR MR. BOTTS: We have just received word that the road commissioners of Willow County, Nebraska, are to hold a meeting on Monday, March seventh, at the county seat, Willow Bend, for the purpose of considering the purchase of a tractor for use in grading and maintaining roads. We want you to attend this meeting and sell them an Earthworm tractor. As they already have our literature and have written us that they are interested, we feel sure that it will be very easy for you to close the deal with them.

Be sure to keep us informed of your progress in your daily reports. We are counting on you.

Very truly,
GILBERT HENDERSON,
Sales Manager.

FARMERS' FRIEND TRACTOR COMPANY
SALESMAN'S DAILY REPORT

DATE: MARCH 7, 1921.
WRITTEN FROM: WILLOW BEND, NEBRASKA.
WRITTEN BY: ALEXANDER BOTTS, SALESMAN.

I received your letter in Omaha and came out here this morning. You were right when you informed me that the Willow County road commissioners were holding a meeting this afternoon for the purpose of considering the purchase of a tractor. I attended this meeting as you suggested. However, you were all wrong when you told me that it would be easy to sell them an Earthworm tractor. As soon as I explain the conditions here you will see that this is a very delicate proposition indeed. It is a very lucky thing that you sent out a man like myself, who is not only a good salesman but who also possesses a clear mind, is a diplomat and understands how to handle an unusually ticklish situation.

I will first relate what took place at the open meeting this afternoon, and I will then describe the dark and devious doings which happened after the meeting.

It was exactly three P.M. when I entered the courthouse and made my way to the room on the second floor where the meeting was to be held. The commissioners—four in number—were already there, and I was pained to observe that there was also a young man representing the Steel Elephant Tractor Company. The meeting was called to order by the chairman, a very large and imposing looking gentleman by the name of George Terwilliger, and the Steel Elephant man was invited to address the commissioners on the subject of his tractor. During this address—which seemed to me very weak and lacking in all the elements of a good sales talk—I improved my time by unobtrusively looking over and sizing up the commissioners.

As I am a very good judge of men, I soon saw that three of the members were very weak indeed, and, that Mr. George Terwilliger was the boss bull of the herd. He is a man of very commanding presence, being over six feet tall and apparently weighing considerably more than two hundred pounds. In general I like big bozos such as this, but as soon as I saw Mr. Terwilliger I knew that I was not going to like him. For he had in his face a low, shifty, underhanded look that to a person of my straightforward nature was most disagreeable. Furthermore, his voice had an ugly nasal twang which grated most raucously on my sensitive eardrums. In spite of his unpleasant personality, however, he seemed to have the other members of the board pretty well buffaloed; he did practically all the questioning of the statements of the Steel Elephant man, and the rest of the commissioners deferred to him on all points.

After the somewhat feeble effort of the Steel Elephant salesman was completed, I arose and delivered a sales talk which was—if I do say so myself—even better than my usual performance along this line. Before the meeting, one of the commissioners had showed me a map of the county, had told me the number of miles of road to be maintained and had outlined a certain amount of new construction which they intended to do. At once I saw that this large and prosperous county had work enough for three or four tractors instead of the one they were thinking of buying.

And when I was told that the county had fifteen thousand dollars in the treasury my mind was made up.

When I came to address the meeting I analyzed the county's needs in a masterly and thoroughly scientific manner, proving that—for the purpose of moving dirt in their new construction, for maintaining their old roads, for snowplow work in the winter and for miscellaneous other work—they would need one ten-ton Earthworm tractor at six thousand dollars and two five-tons at four thousand dollars each. By a strange coincidence, this added up to fourteen thousand dollars and left them one thousand dollars—plus whatever they could borrow—for various new traders, snowplows and other machinery which they would have to purchase.

I spent practically no time at all in pointing out the advantage of tractors over the horses which the county had been using heretofore. I could see that the commissioners were completely disgusted with their old-fashioned methods of doing work and were already sold on the idea of getting machinery. I did, however, touch lightly but very skillfully upon the subject of the Steel Elephant tractor. I stated very positively that I was not there to knock the other man's tractor, but I could not help alluding in a casual way to all the trouble they had had with Steel Elephants in Minneapolis, Kansas City and several counties of Iowa. I assured them that the Elephant was a good machine and that I was at a loss to understand why three of them—which had been sold to a contractor in Council Bluffs last year—were now deposited in a junk yard on the outskirts of that city.

I further stated that all the important financial writers in the country believed that the Steel Elephant Company was on the verge of bankruptcy, and I pointed out that all Steel Elephants would soon be orphaned tractors, for which the unfortunate owners could get neither service nor parts.

I then recited, with most convincing effect, my fourteen reasons why the Earthworm tractor is the best in the world, and concluded by giving some interesting statistics—which I made up as I went along—proving that the Earthworm is selling three times as rapidly as all other tractors combined.

As I sat down and observed the interest in the faces of everybody—and especially in the face of the big but somewhat disagreeable guy called Terwilliger—and as I considered the feeble and unconvincing chirpings of the Steel Elephant representative, I could see no valid reason why this sale should not go through.

At half-past four the meeting adjourned, it being agreed that there would be another meeting at three P.M. the day after tomorrow—

March 9th—at which time they would make their final decision and sign up the orders for any tractors they decided to get. After handing out a copious supply of our beautifully illustrated folders and advertising material, I went back in a most optimistic frame of mind to the local hotel, which used to be called the Hopkins House, but which has recently been repainted and the name changed to Ye Olde Willow Inne.

So far everything had seemed perfectly straightforward and aboveboard, but I was soon to learn that appearances are often deceitful. Just before supper, as I was sitting in my room on the second floor of Ye Olde Willow Inne, a knock sounded on the door.

"Come in," I said. The door opened and Mr. George Terwilliger entered. "How do you do, Mr. Terwilliger?" I exclaimed, springing forward very cordially. "I am indeed honored. Take a chair. Make yourself comfortable. Have a cigar."

I have repeated my exact words, so that you can see I am still the live wire that I have always been. I am always eager—even in the case of a prospect who is personally distasteful to me—to show the greatest cordiality and good fellowship, and thus to break down, by kindness rather than force, whatever sales resistance I may chance to encounter.

"It's a nice day," I remarked as Mr. Terwilliger sat down and took my cigar.

"Yes," he said in his nasal yet oily voice.

"It's a nice little town you have here," I said, "and a most excellent hotel."

Note: I will admit that these last two statements were not entirely in accordance with the facts, but I always aim to be the polished diplomat even at the expense of absolute scientific accuracy.

"Yes," he said, "it is a nice little town; and the hotel is good enough for the people that usually stay here."

My quick mind at once detected that this remark might be a dirty crack at myself, and I was tempted to come back with some brilliant repartee that would have put this yokel in his place. But brilliant repartee, when directed against a prospect, is not good salesmanship, so I merely laughed good-humoredly.

For a moment or two my caller tapped and drummed nervously on the table with his large and powerful hands.

"That Earthworm tractor," he remarked after a while, "seems to be a pretty good machine."

"You bet," I said. "It is the finest machine in the world. You can rest assured that a man of my sales ability would not be wasting his time in selling anything but the very best."

"I suppose," he asked, "that it would mean a lot to you if we bought what you suggested—one ten-ton and two five-tons?"

I stated very positively that I was not there to knock the other man's tractor, but—

"It would mean a lot to me personally," I replied, "and it would mean a lot to the company."

"Just exactly how much would it mean?" he asked, looking at me with what I can best describe as an evil gleam in his eye.

"I don't think I get the exact idea," I said.

At this point Mr. Terwilliger drew his chair very close to mine, looked about furtively, and lowered his voice to a whisper:

"How much would it mean to you? Would it mean as much as a thousand dollars?"

Even when whispering, his voice had that same nasty nasal twang.

"You can speak up, Mr. Terwilliger," I said. "This hotel is well built. The walls are thick and there is no chance of our being overheard. I don't think I quite understand," I went on. "Just what is it you are getting at?"

"Don't try to kid me," said Mr. Terwilliger. "You are a smart young feller, and you ought to be beginning to see what you got to do if you want to put through this deal."

"What I have got to do," I said, "is to convince the board that it is to their best interest to buy one ten-ton and two five-ton Earthworm tractors."

"Exactly! You got to convince the board—which means you got to convince me. Them other guys are nothing but a bunch of sheep, and whatever I say, they do. You must have seen that at the meeting this afternoon."

"Just exactly how much would it mean?" he asked, looking at me with what I can best describe as an evil gleam in his eye.

"Yes," I said, "I guess you are right on that point."

"You are an intelligent man," said Mr. Terwilliger. "Now, as a road commissioner, I am the servant of the people of this county."

"Exactly so."

"It is my duty to see that the county gets good machinery."

"Right."

"And I will never forget my duty. First, last and all the time, I am looking out for the best interests of the voters and taxpayers of Willow County. But on the other hand, I got to consider myself and my wife and children. A man's first duty is to his wife and children."

"That certainly sounds reasonable," I said.

"Yes, sir, I am a farmer and I am not rich. There is a big mortgage on my farm; I got a thousand-dollar payment coming due next month, and I ain't got the money to meet it."

"That certainly is tough," I said.

"It sure is. If I don't raise the money I am going to lose my farm, and I don't like the idea."

"Naturally."

"If I lose my farm," went on Mr. Terwilliger, "I would be down and out; me and my family would be thrown on the county, and it would cost the taxpayers far more than a thousand dollars to take care of us in our declining years. So I figure that my public duty to myself, to my wife, to my three darling children and to Willow County demands that I retain possession of my farm and continue as an honest, self-respecting and self-supporting citizen."

"Yes," I agreed, "it would certainly be a shame if you lost your property."

"I am glad to see," he answered, "that you look at it in a reasonable way. Now it has occurred to me," he went on, "that if you sell us three tractors all at once you ought to be willing to knock off a little something for such a large sale. And as long as I am the only man on the county board that has been intelligent enough to think of this, there is no reason why I should not get the benefit of it. And as long as I am the man that controls the board, it is to your interest to keep on the good side of me. I can promise you that any arrangement we make will be kept secret, so that you won't be bothered by other people asking for the same sort of pure cut. You don't even have to give me a check for the money. All you got to do is to slip me a thousand dollars in ten-dollar bills sometime before the meeting and I will see that the board signs up for three tractors at the full price of fourteen thousand dollars."

Mr. Terwilliger stopped talking and looked at me very intently. As I am a good judge of men, and as my mind is very quick in sizing up character, I was beginning to see that Mr. Terwilliger was not what I would call entirely honest. In fact, I had a distinct feeling that the dark and sinister specter of corruption was hovering over the fair state of Nebraska.

My first impulse was to inform Mr. Terwilliger that he was a dirty skunk. But it occurred to me that he was so large and powerful that it would be unwise for me to be quite so free in my language. Accordingly I modified my remarks.

"I understand your views in this matter exactly, Mr. Terwilliger," I said. "And as I am very anxious to sell you these tractors, I would be most happy to oblige you. Unfortunately, however, I have no authority to cut the regular price on the tractors or to give a rebate such as you suggest."

"That certainly is too bad," said Mr. Terwilliger. "I wanted to get your make of machine, because I think it is the best; but you admitted yourself this afternoon that the Steel Elephant is also a good machine."

"When I said it was a good machine," I replied, "I meant it only in a figurative sense. You must have understood that. In comparison with a broken-down wheelbarrow or a cuckoo clock that has been through a fire, the Steel Elephant may be a good machine, but not in comparison with an Earthworm tractor."

"That may be so and it may not," said Mr. Terwilliger. "But you can be sure of one thing anyway: If you won't play ball with me, I will have to go and talk to that other guy. I understand he has a room in this same hotel on the floors above."

Having said these words, Mr. Terwilliger got up and started for the door.

At this point I will pause to point out that I was now placed in a very difficult situation. At first there didn't seem to be any way of doing business with Mr. Terwilliger without paying him the thousand berries, and this was something I did not want to do—first, because it would be a dishonest proceeding; secondly, because I would hate to see the Farmers' Friend Tractor Company skinned out of the money; and thirdly, because I would hate even worse to see this big bum walk off with so much swag.

On the other hand, it didn't seem right to turn down his offer completely, because that would mean that he would probably get his thousand out of the Steel Elephant man; besides which I would lose the sale and the taxpayers of Willow County would be defrauded into buying Steel Elephant tractors, of which—as we all know—there is no more pathetic a bunch of junk to be found anywhere in this bright and glorious land. I considered denouncing Mr. Terwilliger to the other commissioners or to the public prosecutor, but I realized that it would be useless. It would be his word against mine, and I could prove nothing.

"Wait a minute, Mr. Terwilliger," I said. "As I told you, it is true that I have no authority to put through this deal. But it happens that Mr. Gilbert Henderson, the sales manager of our company, will be in Omaha tomorrow. I will get him here sometime before the meeting, and I am sure if you put the proposition up to him as reasonably as you have to me, he will be able to do something about it."

Note: It will be understood, of course, that my statement regarding Mr. Gilbert Henderson was pure hooey and was made only in order to give me more time to think up some course of action.

Mr. Terwilliger, while considering my proposition, looked at me most suspiciously. But finally, as he could get nothing better out of me, he decided to do as I suggested.

"Very well," he said, "I will stop here day after tomorrow morning—that will be the morning before the meeting. You can tell your Mr. Henderson to bring along the money in ten-dollar bills. If he hands it over to me fair and square I can promise you that the board, at the meeting in the afternoon, will sign an order for the three tractors which you have recommended."

"Very good, Mr. Terwilliger," I said. "I have no doubt but that matters can be arranged." We shook hands and he left.

A few minutes later I went downstairs and ate a hearty supper. It has always been one of my good points that, no matter how worried I may be, I can always eat. After supper I engaged in a casual conversation with the hotel manager; and very skillfully, and without seeming inquisitive, I led the conversation around to Mr. George Terwilliger. The manager was most talkative; and I was surprised and shocked to learn that Mr. Terwilliger, although a farmer, is a very rich man. It seems the old bird has a part interest in the local bank; and instead of being weighed down by a mortgage on his own place, it appears that he holds mortgages on various pieces of property all over the county. It also appears that the three children are entirely mythical and were introduced into the story by Mr. Terwilliger for purely sentimental reasons. The hotel manager further stated that Mr. Terwilliger had only recently been elected to the board of road commissioners. I was very much interested to hear this last statement, because it cleared up a problem which had been troubling me all through supper. Until I heard this explanation, it had seemed incredible that, with a man like Mr. Terwilliger in office, there could be as much as fifteen thousand dollars left in the Willow County treasury.

After pumping the hotel manager dry of all information possible, I came up to my room, where I have ever since been occupied in writing this report and meditating on what course of action I should pursue. This faculty of mine for carrying on two different lines of thought at the same time shows that my mind is very similar to the minds of Julius Caesar and Napoleon. As you know, these great men—like myself—could dictate letters at the same time that they were planning out the next day's activities.

It is now ten o'clock, and although this report is finished, I regret to state that I have not yet arrived at any solution whatever for the exceedingly complex problem which is before me. Accordingly, I will mail this report and continue working my brain. You may rest assured that if any solution to this difficult situation can be found it will be discovered by

Yours truly,
ALEXANDER BOTTS,
Salesman.

Farmers' Friend Tractor Company
Salesman's Daily Report

DATE: MARCH 8, 1921.
WRITTEN FROM: YE OLDE WILLOW INNE, WILLOW BEND, NEBRASKA.
WRITTEN BY: ALEXANDER BOTTS, SALESMAN.

It gives me great pleasure to report that I have put in a day of intense and extraordinary activity. I have laid plans for an elaborate strategic campaign. I have enlisted the aid of science in my struggle against the forces of evil and corruption in Willow County, Nebraska. And by the practical application of the laws of acoustics, electricity and mechanics I hope to bring my activities in that region to a successful conclusion.

It was only a few moments after I had mailed my yesterday's report that there came to me the inspiration for the plan which I am now putting into effect. I at once informed the hotel proprietor that I was going on a short trip, but didn't wish to give up my room, as I would be back next day. I further informed him that I wanted him to reserve the room next to mine for a friend who would return with me.

I then took the eleven o'clock eastbound train, reached Omaha at midnight and spent the rest of the night at the Gifford Hotel. In the morning I called on my old friend Willis Jones, who runs a little second-rate radio and phonograph shop on Farnam Street. I have known Willis ever since the days when we were boys together in school. He is an insignificant little runt; and as he lacks the commercial ability with which I am so richly endowed, he has not been very successful in business. However, in his own line, Willis is a real genius. He has read thousands of books, and he knows practically all there is to be known about electricity. Besides this, he is very skillful with his hands, and when it comes to stringing wire and putting together the most complicated sort of a radio hookup he cannot be beat.

After explaining to Willis the exact situation out at Willow Bend, I asked him if he could fix up an apparatus which I had in mind for the purpose of putting a crimp in the plans of Mr. George Terwilliger. Willis replied that he could easily produce such an apparatus and he would be very glad to help me out. Willis has always felt very grateful to me, because on several occasions I have been able to assist him by means of small loans at times when, owing to his lack of business ability, he had got himself into financial difficulties.

I have described Mr. Willis Jones and his abilities at great length in order to bring out one of the reasons why I am of such great value to the Farmers' Friend Tractor Company. I am hired as a salesman, and I am, of course, a good one. But I am far more than that. I am a man of very wide interests, and I have thousands of valuable friends all over the country. Thus, when I run across a problem—such as the present one at Willow Bend—which cannot be solved by mere sales ability, I am able to call in the help of a friend who is an expert in some other line. In this case I am proud to state that the marvelous scientific brain of Willis Jones will soon solve the dilemma in which I have been placed by Mr. Terwilliger's nefarious proposition.

Willis and I and a young fellow who works for him spent practically the whole day in Omaha, chasing up supplies and assembling the various units of the apparatus which we have decided to use. At five o'clock in the afternoon the three of us took the train out of Omaha, and a little after six we arrived at Ye Olde Willow Inne with two trunks and four suitcases. I engaged an extra room for Willis' young assistant and had the trunks and suitcases carried up to the room next to mine, which I had already engaged for Willis. As I think I have explained before, the hotel is well built, the walls are thick, and it is possible to make considerable noise without exciting anyone's curiosity. After supper we locked the doors, removed the apparatus from the trunks and suitcases and installed it in such a way as to do the most good.

The bed in my room was next to the wall which separated the two rooms. Beneath the bed Willis placed a contraption which I think he called a microphone. From this he ran a couple of wires into his own room through a hole which he bored with a brace and bit. In his own room he had an apparatus which looked to me a good deal like a radio set, but which he said was for the purpose of amplifying the somewhat faint sounds which would be picked up by the microphone, or whatever it was. This amplifying machinery was in turn connected to a phonograph with cylindrical records—something like the machines they use to dictate into in some offices where the stenographers don't know shorthand, or perhaps the boss feels no great urge to dictate to them personally.

I admit that I do not exactly understand how all Willis' apparatus works, and my description of it is probably not over-scientific. Besides the stuff I have mentioned there seemed to be a number of batteries and a lot of other junk that I didn't know what it is for. But I have every confidence in Mr. Willis Jones. And, in addition, we have done some testing.

At exactly ten P.M., just after the installation had been completed, I sat in the middle of my room and related in a low voice one of my very best jokes. And two minutes later, in Willis' room, we played the record on the phonograph, and it sounded so clear and so natural and the inimitable humor was reproduced so exactly that we all three of us practically died laughing.

After this conclusive test, Willis and his assistant went to bed, and I have been writing this report.

Tomorrow morning, when Mr. George Terwilliger sticks his ugly face in my door, I am going to introduce Willis as Mr. Gilbert Henderson, sales manager of the Farmers' Friend Tractor Company. As Mr. Terwilliger explains his dastardly schemes, Willis' assistant will be running the machinery in the next room and recording the whole business. After this we will put on a little entertainment by playing the records for Mr. Terwilliger. When he hears how good his voice sounds on this splendid instrument, and when he realizes that we would not object to putting on the same entertainment for the other commissioners or for a judge and jury, we feel that we will have him eating out of our hand.

It is not my intention to be hard on the man. I will merely suggest to him that he put through the order for three Earthworm tractors and also pay Mr. Willis Jones for the time and material he has used in putting on the entertainment. If he does this, we will take no further action against him. If not, he can use his imagination as to what is likely to follow.

I will now go to bed with a happy mind and with the feeling that all is well. In conclusion I wish to call your attention once more to the superior way in which I am handling this proposition. In the past, in affairs of this kind, it has been the practice of police and detectives to install an apparatus which merely transmits the sound of the conversation from the room in which the dirty dealing is going on to some other room where a stenographer takes it down in shorthand. This method furnishes nothing but a copy of the words that are spoken, and it is always possible that a jury might think the stenographer was crooked.

But I have gone much further than this. Through my friendship for Mr. Willis Jones I have been able to employ the most up-to-date methods and the last word in the wonders of science. Tomorrow morning we will have—not a copy of the words of Mr. George Terwilliger but the words themselves, as clear, as convincing and as lifelike as if the big bum was repeating them himself. It is, indeed, an inspiring thought.

Yours,
ALEXANDER BOTTS,
Salesman.

FARMERS' FRIEND TRACTOR COMPANY
SALESMAN'S DAILY REPORT

DATE: MARCH 9, 1921.
WRITTEN FROM: WILLOW BEND, NEBRASKA.
WRITTEN BY: ALEXANDER BOTTS, SALESMAN.

Well, we have tried our great scientific experiment on Mr. George Terwilliger. And as the results were, in some particulars, slightly different from what I had anticipated, I feel that it is my duty to give a complete and full account of everything that happened.

It was ten o'clock this morning when a knock sounded upon the door of my room in Ye Olde Willow Inne. I at once opened the door and admitted Mr. George Terwilliger. After welcoming him with as much cordiality as I could possibly show to any such filthy swine, I introduced him to Willis Jones.

"Mr. Terwilliger," I said, "I want to make you acquainted with Mr. Gilbert Henderson, the sales manager of the Farmers' Friend Tractor Company. Mr. Henderson, I wish to make you acquainted with Mr. George Terwilliger, the principal member of the board of road commissioners of Willow County."

Willis, in spite of the fact that he is such an insignificant looking little runt and has no commercial ability, spoke up very brisk and businesslike and gave a very good imitation of an important business executive.

"I am very glad, indeed, to meet you," said the alleged Mr. Gilbert Henderson. "At Mr. Botts' suggestion, I have brought along a thousand dollars in ten-dollar bills, and I understand that you have a proposition to make regarding a deal which you think will be of great benefit to both of us."

"Well, as far as that goes," said Mr. George Terwilliger in his ugly, nasal voice, "I have already explained everything to Mr. Botts here, I suppose he has told you the whole story, so I do not see that there is any use in my going over it again."

"Yes, there is," said the supposed Mr. Gilbert Henderson. "I am a hardheaded, skeptical business man, and I cannot afford to pay out a whole thousand dollars unless I am convinced that I will get something in return for it. But, of course, if you do not want to discuss the matter, we will just call it off and I will put the money back in the bank."

"No need to do that," said Mr. Terwilliger. "My proposition is very simple." And in a low but very clear voice he proceeded to tell the pretended Mr. Gilbert Henderson the same story which he had told me the day before, and which I related in my yesterday's report. The proposition that he made was exactly the same, and his talk was embellished with the same picturesque details, even including the three darling children. During the whole recital I remained seated in a corner of the room by the window, gloating inwardly as I pictured in my mind the scene in the next room, where Willis' assistant had been stationed to run the little phonograph and record the nefarious words which were issuing from the ugly face of this unworthy commissioner.

When Mr. Terwilliger had finished, the false Mr. Gilbert Henderson stated that he was well satisfied, and would step into the other room to get the money. According to our plan he was supposed to come back at once with the phonograph and records and put on the little entertainment. But this is not what happened.

When the bogus Mr. Henderson returned after a couple of minutes, he arrived empty-handed.

"I now have the money in my pocket, Mr. Terwilliger," he stated, "but there are one or two points in your proposal that are not quite clear to me. I hate to bother you too much, but I must insist on getting everything straight, and I would appreciate it very much if you would go over the first part of your remarks once more."

Mr. Terwilliger looked a bit suspicious and impatient, but he finally started his whole series of explanations over again, and the counterfeit Mr. Gilbert Henderson asked so many stupid questions that he had to repeat not only the first part of his yarn but all the rest of it as well, including the same three darling children.

All this time I remained seated in the corner, pretending to look out of the window, and I will admit that I was a good bit disturbed in my mind. I could not understand why the so-called Mr. Gilbert Henderson was making Mr. Terwilliger go through his story twice, and, as I am very quick at sizing up a situation, I had a vague feeling that something was wrong. However, I had great faith in Willis' judgment—in spite of the fact that he is such a poor business man—so, making no sign, I waited to see what would happen.

When the story had been completed for the second time, the make-believe Mr. Gilbert Henderson said "Excuse me" and slipped out of the room. As soon as he had disappeared Mr. Terwilliger turned to me.

"That guy keeps popping in and out of here like a rabbit," he said. "It makes me nervous. Where has he gone now?"

"I don't know," I said, rising to the emergency, "perhaps he has gone to get a drink. He will probably be right back."

And sure enough, a moment or so later he came staggering into the room carrying the small phonograph instrument and a box of records. After setting the phonograph on the table and slipping one of the records into place, he turned to me.

"We are ready to go any time you say."

"What is the big idea?" asked Mr. Terwilliger. "What is this thing?"

At this point I arose and took charge of the proceedings in my usual decisive and efficient manner.

"Mr. Terwilliger," I said pleasantly, "the apparatus which you see before you is one of the greatest wonders of science. It has been brought here by this gentleman beside me. He is not, as you have supposed, Mr. Gilbert Henderson, sales manager of the Farmers' Friend Tractor Company. On the contrary, he is Mr. Willis Jones, one of the greatest electrical wizards in this or any land. During the whole of our little conference this morning this marvelous machine has been running in the next room. It has recorded every word that has been spoken, and in a moment we are going to run off the records which we have made so that you can judge for yourself how good they are.

"If you will turn back your mind you will remember that day before yesterday, at the meeting of the road commissioners, I made a talk which proved conclusively that it is to the best interests of Willow County to purchase one ten-ton and two five-ton Earthworm tractors. You will soon see that my possession of these little phonograph records creates a situation which will make the purchase of these tractors by the county very much to your own personal interest as well. If these tractors are ordered, and if you are kind enough to reimburse Mr. Willis Jones for his expenses in putting on this show, we give you our word that no publicity will be given to these records. But if the tractors are not ordered, or if you fail to pay Mr. Jones' reasonable charges, or if you try any funny business whatsoever, it is probable that we will be playing these records very soon for the amusement of judge and jury.... Willis," I concluded, "start the show!"

During this speech Mr. Terwilliger had remained seated in his chair, with his mouth open and with an expression of stupid and helpless amazement all over his ugly face.

He maintained the same attitude while Willis played off all the records—four in number—and they certainly were swell. Every tone, every inflection and every twang of Mr. Terwilliger's unpleasant voice was as clear and distinct as if he were talking himself. And there was no mistaking that voice; there is only one like it in the whole world.

I felt a warm glow of satisfaction as I considered the effect it must be producing on Mr. Terwilliger. For he could not help realizing we had all the evidence needed to convict him of soliciting a bribe. The four records gave his whole proposition from beginning to end, including, of course, the three darling children I had come to know so well.

It was not until the last record was finished that Mr. Terwilliger spoke.

"Is that all of them?" he asked.

"That is all," said Willis Jones.

"And," I added cheerfully, "it seems to me it's enough"

"Yes," agreed Mr. Terwilliger, "it's enough."

And then he proceeded to pull off what I can only describe as a very dirty trick indeed. In fact, his actions were so completely dishonorable and unexpected that we had no chance of counteracting them. In the twinkling of an eye, and without any warning whatsoever, he sprang to his feet, grabbed his chair, whirled it once around his head, and brought it down with a terrific crash on top of the phonograph. Willis and I both sprang forward to stop him from completely destroying this valuable apparatus.

As I have previously explained, Mr. Terwilliger is a very large and powerful man, while I am of slender build, and poor Willis is a mere shrimp. The big brute gave Willis a shove that sent him completely across the room to the door and scared him so much that he immediately opened the door and escaped into the hall. Mr. Terwilliger then gave me a poke in the nose that caused me to stagger backward and sit down rather abruptly in the corner of the room. For some moments I was so dizzy that I didn't know exactly what was going on, but I was dimly aware of a long succession of sounds as of something being broken, split, cracked and busted to pieces. When my senses finally cleared and I opened my eyes, I saw the huge and brutal form of Mr. Terwilliger standing in the center of the room beside

He sprang to his feet, grabbed his chair, whirled it once around his head, and brought it down with a terrific crash on top of the phonograph.

a pathetic mass of wreckage which had once been a phonograph. Mr. Terwilliger was busily stamping on a quantity of granular material which I rightly surmised was all that remained of the four splendid records. After two or three minutes of this exercise he turned to me.

"I guess that will pretty well knock the gizzard out of your little entertainment," he remarked.

"Not at all, Mr. Terwilliger, not at all," came a cheerful voice from the doorway. I looked up and saw that little Willis Jones had just come back in the room. "I am glad to report," he continued, smiling, "that I have sent my assistant to a safe hiding place with the other set of records."

"What?" said Mr. Terwilliger. "Is there another set?"

"Certainly," said Willis. "What did you think we made you tell your story twice for?"

"That's a fact," said Mr. Terwilliger stupidly, "I did tell it twice."

"Sure you did," said Willis, "and we recorded it twice. Naturally, we would not have brought one set in here where you could get your claws on it unless we had another set to fall back on."

"Where are those other records?" demanded Mr. Terwilliger threateningly.

"Just calm yourself," said Willis. "They are safely outside this hotel. It is useless for you to try any more rough stuff, because you will never get your hands on them. They are going to be kept in the safe of the Farmers' Friend Tractor Company. As long as you behave yourself, nobody here will ever know that they exist; but if you try to pull any dirty work we are liable to do almost anything with them."

"Yes, Mr. Terwilliger," I said, suddenly rising up to take command of the situation, "you are licked, and you are sensible enough to know it. You are now at liberty to take your leave, and I trust that at the meeting this afternoon you will act wisely and sensibly."

"And how do I know that you won't spring those records anyway, no matter what I do?"

"You don't," I said very cheerfully. "All you know is that if you behave yourself from now on you have a good chance to get off without being hurt. If you don't behave yourself, however, you can be absolutely sure that our records will make a loud and very disagreeable noise in this county. You are at liberty to choose any course of action you wish."

I have repeated the exact words which I used, so that you can see the masterly way in which I brought this dangerous character to terms. And it gives me great pleasure to report that, after listening to my powerful and convincing language, Mr. Terwilliger was wise enough to choose a course of action favorable to the best interests of Willow County in general and to himself in particular. He took us down to the bank and paid Willis two hundred dollars, which completely reimbursed him for the loss of the phonograph and paid him handsomely for the time he put in. At the meeting this afternoon, the commissioners—upon the advice of Mr. Terwilliger—signed orders for one ten-ton and two five-ton Earthworm tractors. These orders I am inclosing with this report. And thus, against almost insuperable obstacles, another brilliant piece of salesmanship has been accomplished by

It gives me great pleasure to report...

Yours truly,
ALEXANDER BOTTS,
Salesman.

P.S.: After the meeting I told Willis that I was indeed proud of myself for having selected such an excellent assistant.

"When you made old Terwilliger tell his story the second time," I said, "I had no idea you were making a second set of records."

"As a matter of fact," said Willis, "that fool assistant of mine went to sleep the first time, so I had to get the story repeated to make any record at all."

"Then how," I asked, "did you get the second set?"

Willis' reply shows that I am even better than I had supposed; for in getting him to help me I had secured not only a real scientist but also a master strategist and a smooth talker.

"There never was any second set," said Willis.

TROUBLE WITH THE EXPENSE ACCOUNT

ILLUSTRATED BY TONY SARG

FARMERS' FRIEND TRACTOR COMPANY
MAKERS OF EARTHWORM TRACTORS
EARTHWORM CITY, ILLINOIS

MAY 11, 1921.

MR. ALEXANDER BOTTS,
WHITESTONE HOTEL,
CHICAGO, ILLINOIS.

DEAR MR. BOTTS: Our business out on the Coast has recently developed to such an extent that we have not enough salesmen to handle it; accordingly it is necessary for us to transfer you temporarily to our California office.

You will proceed at once to the town of Fontella, California, and call on an English ranch owner at that place—Lord Sidney Greenwich—who, we understand, is in the market for a tractor. We want you to make every effort to sell him an Earthworm.

While you are in California you will be under the orders of Mr. J.D. Whitcomb, Western Sales Manager, Farmers' Friend Tractor Company, Harvester Building, San Francisco. We have written Mr. Whitcomb that you are coming; you will send him daily reports of your activities, and he will inform you what you are to do when you have finished this first assignment.

You will send your expense accounts to the Western office for payment. The three hundred dollars' advance expense money which we turned over to you some time ago should more than cover the cost of your journey from Chicago to California.

Very truly,
GILBERT HENDERSON,
Sales Manager.

Farmers' Friend Tractor Company
Salesman's Daily Report

Date: May 15, 1921.
Written from: On Board Train Bound From Chicago, Illinois, to Fontella, California.
Written by: Alexander Botts, Salesman.

This report, which I am mailing to the Western office, will announce my approaching arrival in California. I understand that Mr. Henderson of the Eastern office has already informed you that he is sending me out to call upon Lord Sidney Greenwich at Fontella; and he has also, I trust, told you that I am one of the best salesmen in the entire Eastern territory. However, as I have never before worked in the California district and as you do not know me personally, it is possible that you may have some doubts as to whether I am the type of man to handle such high-grade business as selling machinery to the English nobility. In order to set your mind at rest and to quiet any fears that you may have, I will give you a few facts regarding my personality and my wide experience which indicate that I am the very man for this job.

Although most of my work for the Farmers' Friend Tractor Company has been with vulgar county commissioners and dirt moving contractors, I am a man of sufficient culture and innate refinement to deal also with men of the highest mental and aesthetic attainments. I am not provincial in any sense of the word, for I have traveled very extensively in Europe. And although this traveling was done in the capacity of cook for a battery of field artillery in the A.E.F. I was able, because of my superior intelligence, to make use of every opportunity to improve my mind. The insight into French character and culture which I acquired from a couple of girls whom I met in Bar-le-Duc has always been a matter of great satisfaction to me. And the urbane polish and *je ne sais quoi* which resulted from several trips to Paris have made me a real man of the world, capable of conversing with anyone—of high or low degree—on terms of complete equality. I will admit that in spite of my familiarity with high society in France, I have never had much contact with the upper classes of England. This does not worry me, however, because I have had a splendid opportunity for observing the habits and customs of these splendid people through my reading of novels and through observing them as they are represented in the moving pictures. Dukes, earls, counts and princes have no terrors for me. I know exactly how to act with these people, how to talk to them, and what to wear when calling upon them.

In order to reassure you on this point and to show you I am right on to my job, first, last and all the time, I will explain what I have done by way of preparation for my visit to His Grace, Lord Sidney Greenwich. Before leaving Chicago I purchased for myself a black cutaway coat with vest to match, a pair of gray striped trousers, an Ascot tie with scarfpin, patent leather shoes, gray spats, gray gloves, a derby hat with stylish low crown and a very good-looking Malacca walking stick.

I have described my preparations at great length so that you can see how necessary they are, and so that you can cheerfully OK my expense account—which I enclose, and which includes $104.20 which I had to pay for these clothes and accessories. I wish to point out that this amount is very low considering the high quality and elegant appearance of the outfit. At every stage of my purchasing I had the interests of the company at heart. I beat down the storekeeper ten dollars on the price of the coat, vest and pants. And you will note that the diamond stick pin cost only ninety-eight cents, in spite of the fact that it looks just as good as a real one.

I further wish to advise that the expenses due to my various purchases and to the cost of my trip have reduced my supply of ready cash to a dangerously low ebb. Consequently I would like you, as soon as you receive this expense account, to send me the amount thereof, $247.51, by Telegraph to Fontella, California.

The train is due in Fontella this evening. I will spend the night at the hotel. And tomorrow morning, when I go out to see His Excellency, Lord Greenwich, you may rest assured that my appearance and demeanor will be entirely adequate for the occasion.

Before leaving Chicago I purchased for myself a black cutaway coat.

I will mail this letter on the train, and it should reach you at San Francisco tomorrow morning.

Let me urge you once more to waste no time in telegraphing me the money. I will look for it tomorrow evening at the very latest.

> Yours sincerely,
> ALEXANDER BOTTS,
> *Salesman.*

FARMERS' FRIEND TRACTOR COMPANY
SALESMAN'S DAILY REPORT

DATE: MAY 16, 1921.
WRITTEN FROM: FONTELLA, CALIFORNIA.
WRITTEN BY: ALEXANDER BOTTS, SALESMAN.

I have put in a very active day and everything would be going swell except for the fact that up to this time I have failed to receive the money which I so urgently requested you to telegraph me. I am completely at a loss to understand what can be the reason for this delay. I stated very positively in my letter that I wanted this money as soon as possible, and I mailed the letter in plenty of time for you to receive it in the morning. In order that you may see how well I have handled matters here, and in order that you may understand how your failure to send me the money is causing me intense embarrassment and may even jeopardize the success of this whole affair, I will describe my activities in connection with His Honor, Lord Sidney Greenwich, and I will also describe the somewhat peculiar situation which exists this evening.

I arrived last night on schedule and spent the night at the Fontella Hotel. This morning, after an early breakfast, I proceeded to pump the various loafers around the hotel for all the information I could get concerning His Gracious Lordship. I discovered that he had purchased a large ranch about five miles from town a little more than a year ago. He has been living there ever since all by himself. The people I talked with seemed to be very stupid, however, and didn't seem to know or care anything about His Lordship's habits or character, so about all I found out was that he very seldom came to town and as far as they knew was a respectable, decent guy. As I could not learn much more about him in town, I decided to go out and call on him at once.

Accordingly I dressed myself carefully and neatly in the new suit of clothes which I think I have briefly mentioned in my yesterday's report. I then hired a taxi to take me to the ranch.

On the way I stopped at the post office, and was much interested in receiving a letter from Mr. J.D. Whitcomb, Western Sales Manager in San Francisco. This letter, of course, was written before the receipt of my yesterday's report. When I read Mr. Whitcomb's suggestion that it might be advisable to give His Lordship an actual demonstration of what the Earthworm tractor can do on a farm, I at once thought it was a splendid idea. And when I further read that a demonstration ten-ton machine, on its way back from the machinery exhibition at Seattle, had been rerouted so as to stop off at Fontella, I was very much delighted. There is nothing like an actual demonstration to back up and reinforce a good sales talk.

From the post office I had the taxi drive to the freight station, which is on the outskirts of town. I presented the bill of lading, which Mr. Whitcomb had enclosed in his letter, to the agent, and when this gentleman told me that the tractor had already arrived and was in a box car next to the unloading platform, I was very much pleased. My pleasure was somewhat lessened, however, when the agent said that I would have to pay $74.80 freight charges before I could get the machine. Upon counting up all the money I had with me I found I had only $8.20—five dollars of which would be needed to pay the exorbitant charges of the taxicab man for a trip out into the country. I didn't like the idea of this big freight bill staring me in the face when my finances were so depressingly low. But at the time I believed that additional funds would soon be in my hands, so I merely laughed, told the agent that I would be back later, and drove on out to His Lordship's estate.

The ranch consisted of a good many hundreds of acres, most of which seemed to be planted with wheat. There was a large barn with a number of outbuildings and a bunk house and kitchen for the men who worked on the place. A little farther along, set in a pleasant grove of trees, was a cute little bungalow, which the taxi man said was the residence of the big guy himself. We drew up beside the road in front of the bungalow, right behind another car. As I have a quick eye, I at once noted that this other machine was covered with dust, as if it had just completed a long journey. It was large, powerful and expensive looking, and a chauffeur in uniform sat behind the wheel. As soon as I had noted those facts I sprang out of the taxi and advanced toward the house, twirling my cane gracefully as I went. I mounted the steps, crossed the broad veranda, and was just on the point of ringing the front door bell when I became aware of the sound of voices issuing from a partly opened window near the door. I paused—not because I wished to be an eavesdropper but because I felt it would be an intellectual treat to listen to the British carrying on an informal conversation. The first words I heard were spoken by a woman.

"Shut up, you big bum!" she said, in a high, nervous voice. "I didn't come here to argue with you, I came here to tell you what I want."

"Very well," said a man's voice, which sounded somewhat disgusted and weary. "I am listening."

"I ain't going to stand for any nonsense," the woman went on. "I tell you again I am tired of scraping along on the miserable allowance that you give me."

At this point I began to realize that I was not getting as much of an intellectual treat as I had expected. But as the conversation seemed fairly interesting I decided to refrain from ringing the bell for a little while longer.

"I have been sending you a hundred dollars a week," said the man's voice—"which is more than I can afford."

"Don't try to kid me," said the woman. "You got money."

"Not as much as you think."

"You own this ranch, don't you?"

"Yes," said the man, "but there is a heavy mortgage on it. Besides, I fail to see why I should be expected to support you at all. You refuse to live with me."

"Of course I do," said the women, very loud and sneering. "Nobody but a fool would expect me to come out and live in this filthy hole of a ranch."

"Before we were married," said the man, "you told me that you would enjoy nothing more than living on a ranch with me."

"I don't care what I said before we was married. That was like promises before election, and don't mean a thing now. We might as well get down to business. You married me. Whether you like it or not, you are stuck with me. I am a city girl, and I ain't going to live on any ranch—and you can just put that in your soup and inhale it. I am so completely sick of you that I wouldn't live with you anymore, anywhere, anyhow."

"And yet you expect me to support you?"

"What I want out of you," the woman went on, "is enough cash so I can lead my own life in town. I know that you have more money than you let on, so you better be a good sport and cough up. If you will give me five hundred thousand even I will lay off of you from now on; we will arrange for a divorce and everything will be fine and dandy. But if you don't come across, believe me, I will make it hot for you."

"Just how are you going to make it hot for me?" he asked.

"You will find out quick enough," said the woman. "I have talked to a lawyer down in San Francisco and I have found out a whole lot about the law. I have found out things that you don't dream of, you ignorant foreigner. I can bring action for divorce, and after you get socked counsel fees and alimony, and after you have turned over my share of the property and paid your own lawyers, you will be busted higher than a kite. You will wish then that you had been reasonable and done what I told you."

"But I do not possess five hundred thousand dollars," said the man's voice, "and even if I did, I would not give it to you. If you had ever done anything to try to make this marriage a success, it would be different. But you have not. You have no real claim on me."

"You married me, didn't you?"

"Yes, after you got me so drunk I didn't know what I was doing. But you have never lived with me. You have never helped me or been good to me in any way. So I cannot see that you deserve anything from me. I am willing to continue the hundred dollars week just to keep you quiet, but you won't get anything more. That is final."

"All right, you big cheese," said the women. "I suppose I might as well be on my way. But you will hear from me again, and you will hear from me soon."

At this point it suddenly occurred to me that I was listening to a conversation not intended for my ears. And as I didn't care to spy on anyone, and furthermore, as it seemed that this very talkative woman might soon be coming out the front door and finding me, I decided to withdraw. I slipped down the front steps, ran back to the taxi, and climbed in.

A moment later the front door opened and there came forth a young woman of the kind best described as being very pretty, in a hard sort of way. She was very expensively and fashionably dressed. The chauffeur opened the rear door of the big car, the young lady got in, and the car drove away. As soon as it had disappeared in the distance, I alighted from the taxicab and once more ascended the steps of the bungalow veranda.

In answer to my ring, the door was opened by a good-looking man about thirty-five years of age, dressed in a corduroy suit and flannel shirt.

"I wish to see His Royal Excellency Lord Sidney Greenwich," I said with quiet dignity.

"My name is Sidney Greenwich," he replied. "What is it you wish?" The voice was the same as the one I had heard talking to the peppery lady.

"My name is Alexander Botts," I answered, "representative of the Farmers' Friend Tractor Company, makers of Earthworm tractors. I understand that you are thinking of buying a tractor and I would be glad to talk to you about our machine."

"Quite so," he said. "I have been thinking of buying a tractor, but at the present time I am rather taken up with certain private affairs, so that I am afraid I cannot give you very much time. Would it be possible for you to come some other day?"

"I would rather not do that," I said politely. "Could you not give me just ten minutes?"

"Well, if it won't take longer than that, perhaps I can," he said, and took me into his simple but adequately furnished sitting room.

I immediately launched forth and gave him a brief, snappy and convincing five-minute sales talk, telling him all the facts he needed to know in order to realize that a ten-ton Earthworm is the ideal tractor for him to use on his big wheat ranch. We spent the remaining five minutes in conversation. And it gives me pleasure to report that His Reverend Lordship is a fine fellow and in all ways a regular guy. He is a real member of the nobility, all right, but he told me that as he expects to remain in America permanently, he has dropped his title and prefers to be known simply as Mr. Sidney Greenwich. The last name, it appears, is pronounced Grenidge and not Green-witch, as might be supposed.

Upon my asking him how a guy gets to be a lord anyway, he replied that in his case it was because he happened to be the younger son of a marquis. In order to let him know that I was wise on such things, I stated that I had often heard of a marquis by the name of Queensberry, who I understood was quite a bozo and the Tex Rickard of his day. Upon my asking Lord Greenwich whether he was related to the Marquis of Queensberry, he replied he was not.

During all this conversation I was surprised and even pleased to note that His Gracious Lordship spoke almost as good English as I do myself, and did not attempt to pull off any of that outlandish language which is used by Englishmen in musical comedies. He did not "bah Jove," nor did he at any time address me as "old chappie."

TROUBLE WITH THE EXPENSE ACCOUNT

In regard to the Earthworm tractor, he said that he had already read up on the subject and made inquiries about it, and that it seemed to him the best machine for his purposes. He said, however, that he would prefer to see one in operation before he made up his mind definitely. Accordingly I informed him that I had a tractor down at the freight house, and that I would bring it out that very afternoon or the next day and give him a real demonstration.

As I have a natural sense of courtesy and of the fitness of things, I didn't attempt to kid His Lordship about the way his wife had been razzing him before I came in. In fact, I didn't even mention the fact that I had overheard the little family quarrel. And realizing that he had probably not been telling any lie when he said he was somewhat taken up with personal matters, I tactfully arose at the end of ten minutes and took my departure. As Lord Greenwich showed me out at the front door I am pleased to report that he took occasion to compliment me upon my stylish appearance.

"I hope you will pardon my saying so," he remarked, "but all through our conversation I have been completely fascinated by that suit of clothes which you are wearing."

"Well," I replied modestly, "it is not a bad suit of clothes, if I do say so myself."

"Oh, quite so," said His Lordship. "I should even call it most extraordinary. Where on earth did you get it?"

As may well be imagined, I was greatly pleased to realize that I was so well dressed that an English lord was actually asking me for the name of my tailor. I cheerfully gave him the name and address of the store in Chicago which had outfitted me.

"I hope," I added in a friendly way, "that I have caused you no embarrassment."

"Embarrassment?" he asked.

"Yes," I replied, "I hope you have not been worrying over the fact that I, who am so well dressed, should have surprised you in your simple farm clothes. You may rest assured that I think no less of you on that account, because I realize that when you are really tricked out in your swell court uniform you would make me look positively dingy. So there is no reason for you to be troubled about the matter."

At these words His Lordship smiled most pleasantly. "You have not troubled me; you have not worried me a bit," he said. "In fact, the sight of you coming in here in those clothes has been the one bright spot in a very disagreeable morning."

And with these words he said goodbye and I returned to the taxi and drove back to town. I paid off the taxi driver in front of the telegraph office and went in, confidently expecting that ample funds would have arrived by wire from San Francisco. You may well imagine my disappointment when the operator said that there was nothing for me. It is disconcerting to a man of my temperament to find his activities held up even temporarily for the sake of a few paltry dollars.

In an attempt to keep things moving I went over to the freight office and had a long chat with the agent. I endeavored to persuade him to let me take the tractor at once and pay the freight charges as soon as my money arrives, which I assured him would occur this evening or tomorrow morning. Unfortunately, these freight agents are sometimes hard eggs to talk to, and this one was worse than the average. In spite of logic, flattery, persuasion, and appeals to his better nature he persisted in his bullheaded determination that I must pay the freight before I could have the tractor.

He also stated that it will be impossible for me to unload the tractor tomorrow. It appears that a small circus is arriving in town first thing in the morning, and that it will be necessary to move the boxcar containing the tractor away from the unloading platform. As the circus will be unloading in the morning and loading up again at midnight, it will be necessary to hold the platform for their use all day. In view of those facts, the agent suggested that I unload the tractor at once, and to this I agreed.

After I had gone to the hotel and changed into another suit of clothes, I filled up the machine with gasoline, oil and water, and ran it out of the box car, across the unloading platform, and down the ramp to the ground. As the machine was too big to go into the freight house, I parked it outside and then came back here to the hotel, where I have been writing this report.

After paying for the gasoline and oil, I find that I have only eleven cents left. I will not attempt to disguise the fact that I am feeling very low in my mind. But I am not giving way to despair. I am still hoping that funds may arrive this evening or tomorrow morning. And in the meantime, owing to the fact that this is an American-plan hotel, I will at least be able to eat. As the supper bell is now ringing, I will close.

Yours very truly,
ALEXANDER BOTTS,
Salesman.

P.S.: Later. 7 P.M. A special delivery letter has arrived from Mr. J.D. Whitcomb, Western Sales Manager of the Farmers' Friend Tractor Company. I have just finished reading it. Needless to say, I could hardly believe my eyes when I saw Mr. Whitcomb's opening statement to the effect that wearing apparel is not considered a legitimate part of a salesman's traveling expenses, and that he most emphatically would never pass the charges for what he describes as an "idiotic and totally unnecessary fancy dress costume." The only reply I wish to make to this remark is to refer you back to Lord Greenwich's opinion which I have quoted earlier in this report. As this costume, more than any other one thing, has served to place me on a favorable footing with this discerning nobleman and important prospect, it seems to me that the cost of it has been well justified.

I will pass over Mr. Whitcomb's next statement—to the effect that the balance of my expense account will not be paid until the end of the week, owing to the fact that my report was received on Monday and it is the rule in the Western office to issue checks only on Saturdays. But I cannot help remarking in passing that a rigid adherence to this office rule may tie up my activities in this region so that I may not be able to put through this sale.

In regard to Mr. Whitcomb's closing statement—that, owing to the fact that I have three hundred dollars' advance expense money, I ought to have plenty of money to carry me along—a word or two of explanation is necessary. According to the system in the Eastern territory—and I suppose also in the Western—a salesman is given a certain amount of advance expense money when he first starts out. If he spends, let us say, fifty dollars in traveling expenses during his first week, he sends in an expense account for this amount, and the office promptly sends him a check for fifty dollars, which brings his total amount back to the original sum he started with. This process is repeated each week, and the theory is that each week the salesman's advance expense money is brought back to the original sum—which in my case was three hundred dollars. I have used the word "theory" advisedly, because every salesman knows that in actual practice the original sum has a habit of gradually wasting away. This is exactly what has happened to me during the past year, and I now find that I have three hundred dollars of advance expense money which I have not got.

Of course, if the company at any time wants its three hundred dollars, I can make it up in time out of my salary; but at the present time this is impossible, as I have no salary due me until the first of the next month.

All of this is perfectly normal, and everything would have been all right except for this stupid holding up of my expense account. For the good of the business and in order that I may proceed with my work, I would strongly urge that you send me that money as soon as you receive this report. Until I get the money, my hands are tied. I cannot get the tractor; I cannot put on a demonstration; I cannot make the sale. Therefore let's get down to business: send me the money, and we will argue about it afterward. I will expect to receive the full amount of my expense account by telegraph tomorrow.

Very truly yours,
ALEXANDER BOTTS,
Salesman.

FARMERS' FRIEND TRACTOR COMPANY
SALESMAN'S DAILY REPORT

DATE: MAY 17, 1921.
WRITTEN FROM: FONTELLA, CALIFORNIA.
WRITTEN BY: ALEXANDER BOTTS, SALESMAN.

My report for today will be a very full one. I will first relate the events of last night and today, and then I will have a few remarks to make on the subject of expense accounts, and I trust that these remarks will be given earnest and respectful consideration by the authorities at the Western Sales Office.

After sending in my yesterday's report, in which I stated that my hands were tied until the arrival of additional funds, I thought the situation over very carefully. And I decided that although the hands of any ordinary salesman would indeed be completely tied by any such chain of adverse conditions as I have encountered, it was, nevertheless, a fact that this need not necessarily be the case with an energetic character such as myself.

There is nothing that I hate worse than remaining idle when there is a good possibility of making a sale. And although I was not being properly backed up by the Western office and although I was being stabbed in the back, so to speak, by the holding up of my expense money, I decided that for the sake of my own self-respect and out of loyalty for the Eastern Sales Office which had sent me out here, I must do something. I decided that,

TROUBLE WITH THE EXPENSE ACCOUNT

money or no money, I would hold that demonstration for His Lordship tomorrow morning.

I set my little traveling alarm clock for two A.M. I retired early. At the first tinkle of the alarm I sprang out of bed and shut it off, so that it would disturb no one else. After dressing carefully in my cutaway coat and all the etceteras, I silently and carefully sneaked down to the ground floor. At the bottom of the stairs there was a slight disturbance which except for my admirable presence of mind might have interfered with my plans. I inadvertently stepped on a large cat which had no more sense than to be asleep on the bottom step, and the wretched animal at once started to rush about the hallway, yodeling loudly as it went. But my quick action in slipping out of the front door and speeding up the street got me away without being observed.

The night was dark and overcast, which well suited my purpose. I soon reached the freight station, which as I have said is on the outskirts of town and far from any dwelling houses. The place was entirely deserted, and no one bothered me as I cranked up the motor and drove away in the beautiful big ten-ton Earthworm. As I have a very good sense of direction I had no trouble in finding the road to His Lordship's farm. It was about four o'clock when I arrived, and I sat around smoking cigarettes until the sun came up and various noises issuing from the bungalow indicated that the owner was awake.

I then rang the front door bell and was admitted by His Most Excellent Lordship himself, who was dressed in the same simple farm clothes which he had worn the day before. I announced that I was ready to put on a demonstration of any kind he desired. His Lordship was very polite, and asked me if I would not join him at breakfast—after which we could go out and run the machine. As I had had nothing to eat since the night before, I gratefully accepted, and we had a very good breakfast, prepared by Mr. Greenwich's serving man, consisting of coffee, oatmeal, bacon and eggs, toast and marmalade. During the meal Lord Greenwich acted in the pleasantest and most friendly manner, and I found myself getting to like him more and more. He told me that he was very glad indeed that I had come out, not only because he was interested in tractors but also because having me around tended to distract his mind from his troubles.

I was too tactful to ask him what these troubles were. But having heard his conversation of the day before with his wife, I could make a pretty good guess, and I could well imagine that he might be in a rather low frame of mind.

During the meal Lord Greenwich acted in the pleasantest and most friendly manner.

Before the end of the meal he took occasion to compliment me once more upon my clothes. And in view of the fact that there has been ignorant criticism of the garments which I have seen fit to procure for myself, I will repeat his discerning Lordship's exact words.

"You have no idea," he said, "how those clothes of yours delight me. I can hardly keep my eyes off them."

And that is not all. When he heard that I intended driving the tractor and handling the dirty plows and farm machinery in this same suit, he at once protested.

"It would be a shame," he said, "to risk spoiling such an elegant outfit. Why don't you change into something else? I have plenty of old clothes and I shall be delighted to lend you some."

As His Lordship was most insistent, I gracefully gave in, reflecting that by so doing I would make myself even more popular than ever with my prospect. For it is a well-known principle of psychology that you can often gain a person's good will by gracefully allowing him to do you a favor and then acting as grateful as possible.

Accordingly I permitted His Lordship to take me up to his room and dress me in a complete outfit of his own old clothes—coat, knickers, flannel shirt, felt hat and heavy shoes. And as His Lordship and I are about the same size, the garments fitted me very well.

As soon as this change had been made I took His Lordship out to the tractor and gave him a thorough and splendid demonstration. His Lordship, in spite of the fact that he is a foreigner, appears to have real intelligence. His questions showed that he has a thorough grasp of the principles of automotive engineering. We put in a very interesting and profitable two hours driving about, hooking onto various pieces of farm equipment, plowing, harrowing, and pulling up stumps and otherwise testing out the strength of the machine. During this whole time I felt my liking for this gentleman increasing more and more, and it was with great sadness that I reflected upon the fact that he was so unfortunate as to be married to the disagreeable person I had heard speaking the day before. My sorrow for the sad plight of His Lordship, however, was somewhat mitigated by the thought that I myself have so far escaped a similar misfortune. And I could not help congratulating myself upon the fact that although I am fairly attractive to the opposite sex I have been strong-minded enough to resist all blandishments and have remained a free and independent bachelor.

At the conclusion of the demonstration I parked the tractor in front of the bungalow, and His Lordship went inside to see if there was any

We put in a very interesting and profitable two hours driving about, hooking onto various pieces of farm equipment, plowing, harrowing, and pulling up stumps and otherwise testing out the strength of the machine.

news from a man whom he said he was expecting from San Francisco. I told him I would follow as soon as I had tightened up a leak which had developed in the gasoline supply line. His Lordship had not yet stated whether he would buy the tractor, and it was my intention, as soon as I should have rejoined him in the house, to take out my order blanks and enter upon the final stages of my selling campaign. As His Lordship had apparently been much pleased with the tractor, I had little doubt of being able to get his name on the dotted line within a very short space of time.

However, at this point an entirely new complication was suddenly injected into the proceedings. As I was putting away the wrench after having tightened the union nut on the gasoline line, I happened to glance down the road and saw an automobile approaching from the direction of the town of Fontella. It was a touring car. There were two men in it, and as it approached I had, for some reason or other, a premonition that all was not well. Ever since I had taken the tractor away from the freight house I had had a vague feeling in my subconscious mind that I might be getting myself into trouble. Of course, I had not really stolen the tractor, for it had been consigned to me. Furthermore, I was not even trying to beat the railway company out of the freight bill, because I intended to pay it as soon as I got the money. But the freight agent was a stubborn old horse, and it was perfectly possible that in case he noticed the disappearance of the tractor, he might get the law after me. And if he did this, it was possible that I would find myself technically in the wrong and perhaps even actually in jail. In order to put through this important sale I had cheerfully assumed this risk, but I will admit that I had been a little uneasy about it. And now that I saw these two men approaching from town it suddenly flashed across my mind that they might be officers of the law. And if they were, it occurred to me that the simplest course would be to withdraw to some secluded spot and avoid meeting them for the time being. Later on, when I had funds to pay the freight bill—provided the Western office finally did have sense enough to send them on—I would be in a much better position to talk to the constable, sheriff, or whoever it might be. It was too late to get the tractor out of sight, so all I could do was to save myself.

I didn't wish to attract attention by running, so as the automobile approached I started strolling in as careless and casual a manner as I could toward a hedge which started not far from the bungalow and continued a couple of hundred yards to the barn and other outbuildings. If I could once get to this hedge I felt sure that I could run along behind it unobserved and hide in the barn until the excitement blew over. However, the automobile drew up in front of the house before I had quite gained this friendly concealment. I heard one of the men say "There he goes now!" And then he called out in a loud voice, "Hey, there, you! Come back here! We want to talk to you!"

To this command I made no reply, but continued walking somewhat faster toward the end of the hedge.

"Stop!" yelled the man. And glancing over my shoulder I saw both men leap from the car and start after me as fast as they could.

As concealment was no longer possible, I immediately lit out for the barn at top speed, with the two men following right after, hollering and yelling and even threatening to shoot if I didn't stop. Courageously disregarding these threats, I kept on. And as I am naturally fleet of foot and always in good condition, I gained the barn with a fairly good lead. I at once climbed into the upper story, or haymow, with the intention of hiding under the hay, but to my dismay I discovered that this would be impossible, for the reason that there was no hay. Apparently the last year's crop was all used up and this year's crop had not yet been cut. It was too late to go down the ladder, as the two men were already climbing it, but I had just time to scramble out of a window onto a shed roof, from which I jumped to the ground and started off across the open fields. The two man came right along after me, but even so, my ability as a runner would probably have got me away, except for a sudden accident. As I ran, my foot went into a small hole that had been dug by a gopher, wood chuck, prairie dog, or some such animal, and I landed on my face in the dirt. Before I could get up again the two men were upon me.

"Are you Lord Sidney Greenwich?" asked the man who had been doing all the hollering.

This question puzzled me somewhat, in view of the fact that it was me and not His Lordship who had swiped the tractor. However, I didn't wish to be of any assistance to these vulgar persons, and I made use of an expression which I have often read in accounts of famous trials.

"I refuse to answer," I said as soon as I could catch my breath, "on the ground that it might incriminate me."

The two men looked at each other, and the first one remarked, "Of course he won't answer; he is trying to dodge service."

"We should worry," said the other man. "It's him all right. Look here!" With these words he exhibited the hat which Lord Greenwich had lent me and which had come off when I fell down. Inside the hat was the

As I ran, my foot went into a small hole that had been dug by a gopher.

trademark of a London manufacturer and the initials S.G. The first man looked at the hat, then drew from his pocket a large folded paper, which he handed to me.

"You certainly gave us a run for our money," he remarked, "but we got you all right." Then turning to the other man, he said, "Come on, Jim, let's go," and the two of them walked back toward their car while I remained seated on the ground resting myself and reading the paper which they had presented to me.

The paper was addressed to Sidney Greenwich. It was very long and was written in idiotic legal language. I could not understand more than half of it, but I gathered that it was an order from some judge which prohibited or enjoined Mr. Greenwich from selling or disposing of any of his property, either real or personal, pending the appointment of a receiver, who, it appeared, would take charge of everything he owned on account of a divorce action which was being brought by his wife.

As I finished reading this paper, I looked across the fields and saw that the two men had got back to their automobile and were driving toward

town. I then arose and walked back to the bungalow. As I did so I meditated on this curious incident; and my natural intelligence told me that these men had served their little paper on the wrong guy—which might turn out to be a fortunate circumstance.

When I arrived at the bungalow His Lordship was on the front porch. Beside him stood another gentleman, who had apparently just arrived in a large car which was parked in front of the house.

"Well, well, Mr. Botts," said His Lordship as I came up the steps, "you are truly a most extraordinary person. I have been watching you from the window with the greatest interest. I can hardly wait to find out who your two friends may be, and what is the meaning of this new game of hare and hounds, or whatever it is. By the way," he added, "this is Mr. Hendricks, my lawyer, who has just arrived from San Francisco."

I shook hands with Mr. Hendricks and we then entered the living room. At once I started on a masterly course of action.

"Mr. Greenwich," I said, addressing His Lordship, "I have a confession to make. Yesterday morning I inadvertently overheard part of the conversation you had with your wife. I thus know more about your affairs than you suppose. I feel, however, that you will forgive me for my accidental eavesdropping after you have heard a few remarks on the subject which I purpose to make. I have a few questions which I would like to ask Mr. Hendricks."

"Very well," said Mr. Hendricks, "proceed."

"Suppose," I said, "that Mr. Greenwich's wife brought suit for divorce. Would it not be possible for her to get out some sort of an injunction which would completely tie up Mr. Greenwich's property pending the settlement of the suit?"

"That would probably be the first thing that she would do," answered Mr. Hendricks.

"Exactly so," I said. "Now let us suppose that this injunction was brought out here by two men in an automobile. Let us suppose that some very good friend of Mr. Greenwich happened to be out in the front yard dressed in a suit of Mr. Greenwich's clothes; and let us suppose further that when he saw those two men approaching he immediately suspected what might be their errand and drew them off by running away across the fields. In such a case it is possible that the two men might run after the friend and give him the injunction in spite of the fact that the friend refused to tell his name. Let us suppose, also, that this friend burned up the injunction without telling either Mr. Greenwich or his lawyer about it. The question I wish to ask is this. In such a case, would it not be true that the injunction would have no legal force on Mr. Greenwich?"

Having said these words, I drew the paper from my pocket and threw it into the open fire which was burning in the room. Mr. Hendricks watched the paper go up in smoke, and then he looked at me for some time very keenly.

"In such a case as you have described," he said finally, "it is my opinion that the injunction would have no legal force."

"Good," I said. "Let us do a little more supposing. If the friend had not acted as he did, it is almost certain that the two men would have come to the house and served the injunction. If this had happened, would Lord Greenwich have been any worse off than he is now?"

Mr. Hendricks turned to His Lordship. "Have you any objection to my discussing your affairs with this gentleman?" he asked.

"None at all," said His Lordship.

"In case my client had been served with such a paper," said Mr. Hendricks, "it is not exaggerating to say that it would have been a very serious

TROUBLE WITH THE EXPENSE ACCOUNT

matter. Mr. Greenwich has a large quantity of wheat which he has been holding for a rise in the market, and which he is now ready to sell. He is planning to use the money from this wheat to purchase various supplies and new machinery which are essential to the farming operations, and to make various needed improvements. The service of an injunction such as you described would have held up all these transactions and caused Mr. Greenwich very heavy losses."

"However," I said, "no injunction has been served and neither of you even knows that such a paper exists. I suppose it is possible that Mr. Greenwich may take a little rest in the house for the next few days, where it is not likely that other process servers would find him. And meanwhile I take it that you, Mr. Hendricks, can attend to those various transactions you have mentioned so as to get Mr. Greenwich's affairs into such a shape that an injunction would not be such a serious matter."

"You have guessed exactly right," said Mr. Hendricks.

"How can I ever thank you," said His Lordship, shaking me by the hand, "for what you have done?"

"You can never thank me at all," I answered, "because you do not know anything about what I have done. Remember, I didn't tell you that I had received any paper of any kind. I merely told you that such a thing might have been possible. However," I went on, "I am now about to do you a very great favor indeed."

"What is that?" asked His Lordship.

"In view of the fact that your affairs have not yet been tied up, I am going to give you the opportunity of purchasing some good machinery. You have seen what a splendid tractor the Earthworm is. You have already practically decided to buy one. But my experience tells me that a ranch of this size will need at least two tractors—one ten-ton and one five-ton. If you do not buy them you will always be sorry. Consequently, I am going to write up an order for the two machines right now, so that you can sign it."

After a few moments' discussion, His Lordship saw the justice of my arguments and signed the two orders. And thus everything has worked out splendidly, with one exception. This one exception—the outrageous handling of my expense account—I will take up as soon as I have narrated the other events of the day. I left the ten-ton tractor in Lord Greenwich's possession and he gave me—as full payment for both tractors—his check for ten thousand dollars, which I enclose together with the two orders. At my request Mr. Greenwich had this check certified, so that there will be no trouble in cashing it if His Lordship's property should be tied up.

His Lordship gave me a separate check for the freight bill—$74.80—and after changing back into my good clothes, I bade him a regretful farewell and rode into town with Mr. Hendricks.

At the freight station I found the agent very sore indeed, and he would have had the law on me except that the only policeman in town was busy out at the lot where the circus was preparing for the afternoon performance. For a time this agent acted very hostile, but after I had given him the check for the freight he quieted down.

After finishing this little matter I stopped at the telegraph office, where I found that $143.31 had been wired me from San Francisco. This money will temporarily alleviate my financial distress, but I was deeply pained to observe that the amount was exactly $104.20 less than the $247.51 which I had expected.

This deduction seems to indicate that my expenses in buying that suit of clothes have not been allowed. And against this procedure I wish to register a deep and solemn protest. I might threaten to resign and leave the company flat. I have had offers from the Steel Elephant Company at a higher salary than I am now receiving. But I scorn to resort to such threats, and I will merely put it up to you, as man to man, that this debt must be paid.

I beg you to consider the following points.

1. I am an artist. I am not like ordinary men. In order to work to the best advantage, I must have the sort of clothes which appeal to my artistic sense and create an atmosphere, in my own mind and in the mind of the prospect, which is conducive to the breaking down of sales resistance. If you attempt to cramp my style you will not only be injuring me but you will also be creating a situation which will cause the company to lose many sales.
2. I get results. Nobody can deny that. You expected me to sell one tractor here, and I have sold two. The profit on this additional business is far more than the paltry $104.20 which I spent on that suit. As soon as I cease getting results, you can cease to pay my expense account. But in view of the fact that the results are always good, there is only one course of action possible.

You are going to pay that $104.20. You will send it by telegraph as soon as you receive this letter. I will be waiting here in Fontella to receive it, and I will also await your instructions for my future activities.

In the meantime I will be working on another possible sale. I am going out to see the manager of the circus and try to induce him to buy a small tractor to help load and unload equipment from railroad cars, to pull out any of the wagons or elephants that get stuck in the mud, and do other odd jobs. In conclusion I wish to remind you once more that I am waiting for that money.

Very truly,
ALEXANDER BOTTS,
Salesman.

TELEGRAM
SAN FRANCISCO CALIFORNIA MAY 18 1921
ALEXANDER BOTTS
FONTELLA CALIFORNIA

IN VIEW OF THE FACT THAT YOU SOLD TWO TRACTORS WE ARE WIRING ONE HUNDRED FOUR DOLLARS TWENTY CENTS AS REQUESTED STOP BUT THIS IS NOT A PRECEDENT STOP WHEN YOU CALL ON CIRCUS YOU WILL PERHAPS WANT TO GET YOURSELF A CLOWN SUIT STOP THIS WOULD UNDOUBTEDLY BE MOST APPROPRIATE BUT WE WILL NOT PAY FOR IT STOP THIS IS FINAL

J D WHITCOMB

THE
BIG SALES TALK

ILLUSTRATED BY TONY SARG

FARMERS' FRIEND TRACTOR COMPANY
MAKERS OF EARTHWORM TRACTORS
EARTHWORM CITY, ILLINOIS

NOVEMBER 30, 1921.

MR. ALEXANDER BOTTS,
MCALPIN HOTEL,
NEW YORK CITY, NEW YORK.

DEAR MR. BOTTS: We want you to go to Chipman Falls, Vermont, to take charge of a demonstration which we have arranged to hold next Monday for Mr. Job Chipman, a big lumber operator of that place. Mr. Chipman is considering replacing most of his horses with tractors.

Mr. Chipman wants to use the machines in plowing out his wood roads following snowstorms, and also in hauling sleds loaded with logs along these roads after they have been sprinkled with water to give them an ice surface.

We have already shipped to Chipman Falls a ten-ton Earthworm tractor equipped with snowplow, winter cab, and ice spike grousers; and we are sending our service man, Mr. Samuel Simpson, to operate it.

You will have competition; we understand that the Mammoth Tractor Company will be on hand with one of their machines.

But we have every confidence in your ability as a salesman, and we are relying on you to get Mr. Chipman's order for one or more Earthworms.

Very sincerely,
GILBERT HENDERSON,
Sales Manager.

FARMERS' FRIEND TRACTOR COMPANY
SALESMAN'S DAILY REPORT

DATE: SATURDAY, DECEMBER 3, 1921.
WRITTEN FROM: NEW YORK CITY.
WRITTEN BY: ALEXANDER BOTTS, SALESMAN.

It was indeed a pleasure to receive your letter of November 30, and you may be sure that you are making no mistake in sending me on this important mission.

As you know, I have always been one of the best natural talkers in your whole organization. But I am never satisfied; I am always trying to improve myself. And realizing the fact that I had never had any experience in selling tractors to northern lumbermen, I have spent the past two days—ever since receiving your letter—at the New York Public Library, studying up on all phases of the lumber business as it is practiced in the northern part of the country. It gives me great pleasure to inform you that I have now acquired all the information and knowledge that are necessary in order to adapt my already splendid sales arguments to the particular conditions which I shall meet with when I begin working on Mr. Job Chipman. It's not exaggerating to say that I will start on this selling campaign with a sales talk that is really big. I can hardly wait to get going.

I have engaged a berth on the sleeper leaving Grand Central tomorrow night and arriving at Chipman Falls early Monday morning. You may rest assured that by Monday night Mr. Chipman's order will be safely in the mail and on its way to your office.

Very sincerely yours,
ALEXANDER BOTTS,
Salesman.

Farmers' Friend Tractor Company
Salesman's Daily Report

Date: Monday, December 5, 1921.
Written from: Chipman Falls, Vermont.
Written by: Alexander Botts, Salesman.

Well, I have arrived, and I have put in a very busy day. But I may as well admit that things are not going as good as I had expected. In fact, I have run into a veritable nightmare of unexpected difficulties and misfortunes which are absolutely unparalleled in my experience as a salesman for the Farmers' Friend Tractor Company. I will relate exactly what has happened so you can see that I have done everything I could, but that fate has been against me.

When I first got off the train at seven o'clock this morning I had no hint of impending misfortune. The weather was dark and cloudy and very cold, but there was no sign of snow on the ground. I was sorry that we would be unable to show Mr. Chipman how well the Earthworm can travel through snow, but I was not worried. A demonstration is always a good thing, but the real heart and soul of any selling campaign is the sales talk, and I was confident in my knowledge that in this department of the game I was supreme.

As I walked up to the hotel I noted that the town of Chipman Falls is in a deep narrow valley in the Green Mountains, and that it consists of the railroad station, a small hotel, several dozen houses and a large sawmill. The steep mountain slopes on both sides of the valley are covered with timber. The hotel clerk informed me that all this timber, as well as the sawmill, is the property of Mr. Job Chipman.

Leaving my baggage at the hotel, I walked over to the mill. Just outside the boiler house I found our ten-ton demonstration Earthworm, equipped with its big snowplow, in charge of Mr. Samuel Simpson.

Sam reported that he had unloaded the machine the day before, and he had kept it overnight in the boiler house so that it was nice and warm and all ready to go. He had not seen Mr. Chipman, but the sawmill foreman had told him to be ready to put on a demonstration that morning.

"I understand," said Sam, "that Mr. Chipman wants a machine for snowplow work and for hauling sleds on ice roads. As there is no snow, and consequently no ice roads, I don't see how we can put on much of a demonstration."

"Never mind about that," I said. "We will just run the machine around on the bare ground so the old man can see how it goes. Then I will give him such a vivid description of the way it rolls through the snowdrifts that he will think he has actually seen it. Just wait till you hear me get going."

"Fine!" said Sam. "In the grouser box here I have a set of the latest type ice spike grousers; but as there is no snow or ice, I don't suppose there is any use in putting them on."

"No use at all," I said.

At this point there was a sudden clanking and puffing, and out from the boiler house there came a big sixty-horsepower Mammoth tractor in charge of two mechanics and a salesman. I had never before seen this particular model of tractor. But when I looked it over and realized that it was actually attempting to compete with the Earthworm, I will admit that I had to laugh. For the Mammoth is undoubtedly the most outlandish engineering atrocity which has ever been thrown together and offered for sale to the long-suffering public. It is about ten years behind the times in design, and although it has about the same horsepower as the ten-ton Earthworm, it weighs almost twice as much. It is at least twenty feet long, with two small wheels in front and two enormous bull wheels about nine feet in diameter in the rear. The motor consists of one big horizontal cylinder which works on a flywheel of monumental size and weight. All gears seem to be made of high-grade stove iron. The whole works is heavy, slow, and clumsy, and it needs the proverbial ten-acre lot to turn around. It is, in short, nothing but a mechanical joke when compared to the Earthworm, with its modern four-cylinder motor, its small sized steel gears, its light weight, and its ease in steering and turning around.

After I had contemplated the Mammoth for some moments, the salesman in charge stepped up and introduced himself. It appears his name is Jones. As I looked him over I decided that I was not going to like the man. He had a coarse and rather stupid looking face. He was quite evidently lacking in both ideas and ideals. And he was, I decided, the type of person who makes up for his dearth of intelligence by developing a considerable degree of low cunning, which is totally out of place, of course, in the noble profession of selling tractors.

As I say, I didn't like his looks. But, as I am always polite, I concealed my dislike.

"I suppose," I said affably, "that you have already seen Mr. Chipman and have given him a line of talk designed to prove that the Mammoth is a better machine than the Earthworm."

"I have seen Mr. Chipman," replied Mr. Jones, in a disagreeable, rasping voice, "but I have given him no line of talk. And what is more, I don't intend to."

"No?" I said. "What's the big idea?"

"I intend to put on a demonstration for him," said Mr. Jones. "When a man can give a real good demonstration he does not need any sales talk."

"My boy," I said, "I can see you are new at this selling game; you are young and you have much to learn. If you doubt the value of a good selling argument, you had better stick around when I start in on Mr. Chipman. You will see how a real salesman, by the skillful use of the English language, beats down and completely overcomes the sales resistance that may be present in the mind of any prospect."

"Mr. Chipman is in his office now," said Mr. Jones. "Let's go in right away. Nothing would give me more pleasure than to see you doing your stuff."

"Very well," I said, "come on."

I have repeated my conversation with Mr. Jones so that you can see that I am, as always, frank, kindly, and helpful even to a competitor. When I relate what happened later you will see that Mr. Jones is quite the reverse, being one of the most underhanded, slimy and dishonorable men who ever polluted the tractor business with their presences.

Mr. Jones and I entered the office together. Mr. Chipman turned out to be a tall, bony man, perhaps sixty years of age, with a pleasant, weather beaten face. He had been sitting at a desk, but when we entered he arose and shook hands.

"Mr. Chipman," I said, "my name is Alexander Botts and I represent the Farmers' Friend Tractor Company, makers of Earthworm Tractors, Earthworm City, Illinois. As you already know, we have one of our ten-ton machines outside ready to give you a demonstration any time you want to see it. But before we start I would like to bring to your attention a few facts regarding the tractor and the company which makes it—facts which indicate conclusively that the Earthworm is the only feasible machine for a lumbering enterprise similar to the one that you conduct."

At this point I paused an instant, for I noticed that Mr. Chipman was pushing something across the table in my direction. Looking down, I saw that it was a small slate, such as schoolboys used to use in former times. Attached to it by a short string was a slate pencil. Seeing that I was somewhat surprised and mystified by this procedure, Mr. Chipman opened his mouth, presumably to explain to me what it was all about. But as I had got started so good on my sales talk, I didn't wish to be switched off onto the subject of school slates. Consequently I continued talking as if there had been no interruption.

"When you first see this Earthworm tractor, Mr. Chipman," I said, "I wish you particularly to notice the method of propulsion. Instead of running on four wheels like a wagon"—at this point I glanced meaningfully at Mr. Jones—"the Earthworm goes along exactly like a wartime tank. It runs on two steel tracks, one on each side. Each track is like an endless belt or chain, composed of twenty-nine separate track shoes, which are made of the highest grade of manganese steel, triple-heat treated. The surfaces of these shoes are made perfectly flat so as not to hurt the roads. These flat shoes will, of course, slip when the machine is used on snow or ice; consequently we supply a set of sharp cleats, or grousers, one of which is to be bolted onto each track shoe when the machine is to be used on snow or ice. Each of these grousers takes hold like the claw of a cat, giving the tractor a splendid grip and enabling it to pull tremendous loads through snow, ice, mud, rocks, sand, gravel, clay, or anything else in the world."

At this point I noticed that Mr. Chipman had picked up his silly little schoolboy slate and was holding it out in my direction.

As I paused momentarily he said, "Kindly write what you have to say on this slate."

At first I failed to grasp the meaning of this procedure. Then I heard Mr. Jones, who stood behind me, chuckling in a very vulgar manner.

"You can save your lungs," remarked Mr. Jones, in a disrespectful tone. "The old boss is stone-deaf in one ear and can't hear anything out of the other."

I noticed that Mr. Chipman had picked up his silly little schoolboy slate and was holding it out in my direction.

THE BIG SALES TALK

At these words a sudden chill of horror settled down on my mind. I gave Mr. Chipman a quick shrewd glance. He was pointing to the slate with one hand and to his ear with the other.

"I cannot hear as well as I used to," he was saying. "You will have to write down what you have to say."

As my perceptions are very quick and accurate, it was not long before I realized that Mr. Jones had in his untutored way spoken the exact truth. And like a flash my mind apprehended all the sickening implications of this discovery. I had come striding into this office with my usual masterful air, confident in the knowledge that I had the one essential, the *sine qua non*, of a successful sales campaign—a swell line of sales talk. And now, with a sickening shock, I realized that it was all useless. For what good is a swell line of sales talk when it falls upon deaf ears? What can it accomplish when it is directed as Mr. Jones phrased it—at one ear which is deaf and another which cannot be heard out of?

Behind me I could hear Mr. Jones—the big bum—still chuckling to himself. And well he might. Considering his froglike croak of a voice, his almost total lack of brains and his undoubtedly feeble powers of expression, it was actually to his advantage to have Mr. Chipman deaf. But for me, with my pleasing and musical flow of language, with my wealth of imagination and with my masterful command of the English language, it was little short of calamity.

However, I am not the man to lie down and burst into tears when there is work to do. Grasping eagerly at the slate I started to write. But almost at once I realized the futility of this procedure. Using that stubby little pencil and that miserable scratchy slate, it would have taken me hours to put over even an abridged version of my splendid sales argument. Mr. Chipman would doubtless become bored and walk out before I had even half finished. And even if he were willing to wait and read the whole thing, it would be a tremendous letdown from what I had intended giving him. It would lack the breath of life, the musical tones that charm the ear and the splendid ring of sincerity which has brought conviction to so many doubtful minds.

No, my wonderful sales talk was completely blown up, knocked to pieces and sunk without a trace. But I resolved to die fighting.

"My name is Alexander Botts," I wrote on the slate, in bold characters. "I have an Earthworm tractor outside. If you want to see something good, follow me."

Mr. Chipman read what I had written and smiled cheerfully.

"Fine!" he said, reaching for his coat and hat, which hung on a nail on the wall. "I will come right out, and you men can both do a little demonstrating."

As I helped Mr. Chipman on with his coat I could not help feeling that he seemed to be a very decent sort of chap, bearing up very bravely under his infirmity, and I almost wept at the thought of the treat he was missing through not being able to hear my sales talk. As we walked out, Mr. Jones gave an additional proof that his personality is most obnoxious and exactly the opposite of my own pleasant nature and that of Mr. Chipman.

"Thank you," said Mr. Jones, in his disagreeable voice. "You don't know how much I have appreciated observing the effect you made on Mr. Chipman with your wonderful sales ballyhoo. I am very much obliged."

"You can keep the change." I said with dignity. "If I cannot talk to this gentleman I can get along very well by putting on a demonstration for him."

"That's exactly what I said before we came in," said Mr. Jones. "If a man has a good demonstration he doesn't need any sales talk. I am glad to see that you are learning something from me."

To this wise crack I made no response. As soon as we reached the place where the two tractors were standing I turned to Mr. Chipman.

"I would suggest," I said, "that you take a little ride in our tractor. It is unfortunate that there is no snow and ice to give you a real demonstration,

"Thank you," said Mr. Jones, in his disagreeable voice.

but if you will seat yourself on the comfortable cushion beside the driver and take a little ride, you cannot fail to realize the superiority of the Earthworm tractor. I do not wish to knock any other machine, so I will let you see for yourself that the Earthworm has more flexibility, more speed, more power, and handles far more easily than any other machine in the world. It can go over the roughest sort of round, and it is so light and compact that it steers as easily as a velocipede. It can turn right around in its own tracks. It is truly the latest word in engineering science."

"I am sorry," said Mr. Chipman politely, "but you will have to write that down."

He held out the little slate, which he had brought with him and which he wore suspended by a string from a button of his overcoat. With a sinking heart I took the slate. It is impossible for me to describe how it cramps my style to express myself on a filthy square of slate which squeaks and scratches under the strokes of a pencil.

"Do you want a ride?" I wrote.

"Thank you," said Mr. Chipman, "it will give me the greatest pleasure." He climbed into the cab and sat down, and Sam gave him a swell little ride—up and down the road and out over a very rough field. Sam is certainly a peach of a driver. He handled the machine with real artistry, he changed gears with the greatest smoothness, he went over bumps and depressions with no jar at all, and he turned and twisted the machine about so gracefully and beautifully that Mr. Chipman must have been very favorably impressed.

After this ride was over, Mr. Chipman took a short trip in the Mammoth tractor, which lumbered around over the ground in such a clumsy, slow and awkward manner that I almost began to feel sorry for Mr. Jones. It seemed as if the Mammoth had no chance at all. But Mr. Chipman required further demonstration.

"I want you to bring your tractor down the road," he yelled to me.

The Mammoth machine, carrying Mr. Chipman, Mr. Jones and the two mechanics, started off. Sam and I followed in the Earthworm. About a quarter of a mile down the road we came to a medium-sized pond which was frozen over tight. The Mammoth ran out over the ice to the center of the pond, with Sam and me in the Earthworm riding behind.

"Bring that machine over here," Mr. Chipman shouted to us.

Sam followed his directions, which were given with much shouting and waving of arms, and backed the Earthworm around until it stopped with its rear end almost touching the rear end of the Mammoth. At this point Mr. Chipman got off and proceeded to hitch the two machines together by means of a large chain which was carried on the drawbar of the Mammoth. At once I began to feel that there might be trouble ahead.

"Here, here!" I shouted. "What are you trying to do?"

Mr. Chipman climbed up into the cab of the Earthworm and sat down between Sam and me.

"I need a tractor," he said, "that can pull sleds over ice. That is why I have brought you over onto this pond and hitched your two machines back to back I want to see which one of you can pull the most on the ice. You can start any time you want."

"This won't be any test at all," I said. "You will have to give us time to bolt on our grousers. These smooth tracks are all right for pulling on dirt, but they don't have any more traction on ice than a couple of sled runners." I took a quick look at the Mammoth tractor. "Fortunately," I went on, "that other machine is in just as bad shape as we are. Mr. Jones will have to put on some cleats, or grousers, on those big, smooth drive wheels of his before he can do anything at all. You didn't tell us you were going to bring us out on the ice. But just give me time to put on the grousers and I will enter any pulling contest with any machine in the world."

"Please write it down," said Mr. Chipman, holding out his slate. "I don't hear as well as I used to when I was younger."

I grabbed the slate and was just beginning to write when there came a terrific jerk. The driver of the Mammoth had opened his throttle and thrown in the clutch. At once Sam, in self-defense, did the same thing. I looked around and saw that the chain between the two machines was stretched tight. The big bull drive wheels of the Mammoth were turning round and round, and so were the tracks of the Earthworm. But both the wheels and the tracks were slipping on the ice so that neither of the machines was gaining anything on the other.

At first it looked like a draw. But then the driver of the Mammoth, acting under the orders of the obnoxious Mr. Jones, who sat beside him, pulled one of the dirtiest tricks I have ever seen in all my experience in the tractor business. I saw him reach out beside his seat and pull a great big lever which was connected by long steel rods to a couple of sliding collars on the rear axis. The collars slid along the axle, causing a number of dogs, or pawls, to engage in some teeth cut in the hubs of the big drive wheels. As the drive wheels turned, the motion was carried through these pawls to a number of iron rods which in turn forced a lot of heavy iron spikes out through holes in the smooth iron tires of the big bull wheels. As soon as

THE BIG SALES TALK

these spikes came out they hit into the ice, and all at once the Mammoth had splendid traction.

I had never before seen an arrangement such as this, and I had to admit to myself that it was a rather clever scheme—in fact it is the only feature of the Mammoth tractor which is any good at all. But it was enough to put me in a very awkward position. The Mammoth at once started rolling off across the ice, dragging the Earthworm behind it. Sam was helpless. He could spin the tracks as much as he wanted, but they had absolutely no hold on the ice.

I at once began to protest, shouting and yelling and writing on the silly little slate as fast as I could. But nobody paid any attention. And I was treated to the heartrending experience of sitting in a perfectly good Earthworm tractor and being pulled backward across the ice by that big overgrown bunch of stove iron known as the Mammoth. And that was not all. The mechanic who was driving the Mammoth actually had the discourtesy to go on a circular course, dragging the Earthworm—in spite of everything that poor Sam could do—three times around that pond. All this time Mr. Jones kept waving and smiling at me in a most vulgar and insolent manner. When at last we stopped and the motors of both machines had been shut off, I felt like going over and murdering the whole Mammoth crew; but with a great effort I held onto myself.

"Mr. Chipman," I said, with dignity and restraint, "this is an outrage. This demonstration means nothing at all. You didn't give us time to put on our grousers. It's a dirty trick."

"Write it down," said Mr. Chipman, politely pointing to the slate. As everything I had written during the ride appeared to be illegible, I rubbed the slate clean and started again.

"This demonstration is unfair," I wrote. "You should have given us time to put on the grousers."

"What are grousers?" asked Mr. Chipman.

"Points," I wrote, "like those things on the wheels of the other tractor."

Mr. Chipman looked at the Mammoth and then he looked at the tracks of the Earthworm. His face lit up as if he understood, and I began to think I was making some impression on him. But I was wrong.

"I see what you mean," he said. "But you had just as much time as the other fellow. If he can put points on his machine in two seconds, and you cannot, that just shows that his machine is more convenient than yours, as well as being able to pull better."

As soon as I heard this crazy line of reasoning I rubbed out everything on the slats and started to write a full explanation of the true state of

All this time Mr. Jones kept waving and smiling at me in a most vulgar and insolent manner.

THE BIG SALES TALK

affairs. But I had hardly started before Mr. Chipman looked at his watch and then started climbing down out of the tractor.

"I am very sorry," he said, "but I have an engagement to go down the valley to one of my other sawmills, and I will have to get started right away."

"Wait a minute, wait a minute!" I yelled. And then I wrote on the slate: "When will you be back?"

"Late tonight," said Mr. Chipman

"How about another demonstration tomorrow?" I wrote.

"It is hardly worthwhile," said Mr. Chipman. "The Mammoth seems to be the best machine. But I will not sign any orders until tomorrow night, and if you care to do any more driving I will be glad to watch you."

With these words he nodded a cheerful goodbye and hurried back to the sawmill. A few minutes later I saw him driving away in his car.

While Sam and I were unhooking our Earthworm from the Mammoth, Mr. Jones stood around with the most insulting look I have ever seen on the face of a human being.

"Well, Mr. Botts," he said, with a coarse laugh, "it appears that I was right. If a man can put on a real good demonstration he doesn't need any sales talk."

As Sam and I are both gentlemen, we paid absolutely no attention to this clumsy and useless attempt at humor, but cranked up the Earthworm and drove back to the boiler house. Here I left Sam to put on the grousers in readiness for tomorrow's demonstration while I came over to the hotel, where I borrowed a typewriter from the manager and spent the balance of the morning and all the afternoon in writing up the sales talk which I had hoped to give orally. Since supper I have spent my time on this report.

In summing up the day's events, I will not conceal from you the fact that I am a disappointed man and that things look very black indeed. But if I fail in this enterprise you may be sure that I am going down fighting, with all flags flying. Although I am at my best as a talker and orator, I am not entirely helpless—as you no doubt realize from reading my reports—when it comes to slinging the ink.

The little sales talk which I have prepared for Mr. Chipman consists of thirty-one and a half pages of typewritten material, couched in the choicest English, taking up all points regarding the superiority of the Earthworm tractor and explaining clearly and logically all the reasons why today's so-called demonstration was absolutely and completely worthless as a basis for comparing the two machines. It is perfectly obvious that there can be no scientific value in any demonstration which seems to show that the Earthworm is inferior to any other machine.

I will now retire to bed so that I will be fit for whatever work is necessary tomorrow. I wish, however, to report one hopeful event. Early this afternoon it started snowing. And now, at ten P.M., it is still coming down hard, and there are already several inches of snow on the ground. It is obvious that the more snow we have, the harder the traveling will be and the greater will be the superiority of the Earthworm over the heavy Mammoth.

In conclusion I wish to impress upon you that none of today's misfortunes should be blamed on Mr. Sam Simpson. I consider Sam one of the best mechanics in the employ of the company; he has done everything he could; and if any blame is to be attached to anyone, it should be upon the shoulders of

Your unfortunate salesman,
ALEXANDER BOTTS,
Salesman.

FARMERS' FRIEND TRACTOR COMPANY
SALESMAN'S DAILY REPORT

DATE: TUESDAY, DECEMBER 6, 1921.
WRITTEN FROM: CHIPMAN FALLS, VERMONT.
WRITTEN BY: ALEXANDER BOTTS, SALESMAN.

Never in all my experience as a salesman have I run into so much hard luck concentrated into a single selling campaign. My plans have once more been frustrated by an unexpected and crushing blow, which I will describe later on. This blow was doubly discouraging in view of the fact that it came at a time when I had every reason to suppose that matters were in a slightly more hopeful condition.

When I got up this morning I observed that the snowstorm was over. The weather was clear and cold, and there seemed to be about two feet of snow on the ground. There had been practically no wind and so the snow was level. I had hoped for a regular blizzard with big drifts. Such a condition would have been a wonderful thing for us, because the Earthworm, of course, can go through the deepest snow there is, whereas I knew the big clumsy Mammoth was sure to get stuck in the first big drift it came to.

Although two feet of snow is not enough to stop the Mammoth, I knew it would be slowed up a good bit and that we could therefore make a slightly better showing with the Earthworm.

This slightly better showing, of course, would not be sufficient to overcome the bad impression caused by yesterday's so-called demonstration, so I was relying particularly upon my masterful sales argument in thirty-one and a half typewritten pages, which I had produced at such great labor yesterday. With my precious document buttoned safely into the inside pocket of my overcoat, I sallied forth with Sam immediately after breakfast.

We ran the Earthworm, fully equipped with its grousers and its snowplow, out of the boiler house, where it had spent the night. Mr. Jones and his mechanics brought out the Mammoth immediately afterward. On the front of their machine they had adjusted a snowplow somewhat similar to ours.

A few minutes later Mr. Chipman appeared and explained what he wanted us to do. He pointed out a road which led from the sawmill up the wooded mountain slope on the west side of the valley.

"That is my principal logging road," said Mr. Chipman. "About half a mile up the slope it divides into two roads. You can see the north fork running straight along the face of the mountain yonder." He pointed. "And you can see the south fork zigzagging up around that shoulder. Beyond the shoulder the road is hidden in that little side valley, but you can see it coming into view several miles farther on and zigzagging across the slope of South Mountain, quite high up and a long way off."

We all looked, and sure enough we could see the faint traces of the two roads as they zigzagged along through the woods on the steep mountainside.

"I want to see," he continued, "how well you people can plow snow. You can start any time you want. One of you can take the north fork and the other the south. You can plow out three or four miles apiece and then come back. Set your snowplows fairly high and leave about a foot of snow so we will have a good basis for an ice surface when we get out the water wagons and begin sprinkling. You can start any time you want."

"Fine!" said Sam. "I am on my way right now."

"Wait a minute," I said, and then turned to Mr. Chipman.

"Would you like to ride along in our machine?" I asked.

"You will have to write it down," said Mr. Chipman. I wrote it on the slate.

"No, thank you," said Mr. Chipman.

"All right, Sam," I said, "you can go, I will stay here with Mr. Chipman."

As Sam drove off up the road through the snow I reached into my pocket and brought out the precious paper. I handed it to Mr. Chipman.

"What is this?" he asked.

In reply, I pointed to the title at the top of the first page: One Hundred and One Reasons Why the Earthworm is the Tractor for You.

As I looked at the title of the work upon which I had spent so much careful thought and hard labor the day before, I felt within myself a pardonable glow of pride. But this glow was short-lived. Because at this moment there descended the unexpected and crushing blow which I mentioned at the beginning of this report.

Instead of starting to read, Mr. Chipman handed the paper back to me.

"I am sorry," he said, "but I broke my reading glasses yesterday and they won't be fixed for several days."

To this remark I made no reply. For once in my life I was speechless—which perhaps was just as well, as Mr. Chipman couldn't have heard anything anyway.

"Of course," he continued, "I can read to some extent even without my glasses, but if I tried to go through as much fine print as that it would be too much of a strain on my eyes. Besides, I am more interested in seeing the tractors work than I am in reading about them. And I have promised Mr. Jones I would ride along in his machine this morning."

With these words he climbed aboard the Mammoth tractor, which immediately started off up the road, carrying with it—in addition to Mr. Chipman—the two mechanics and Mr. Jones. And there was I, left standing in the snow with my carefully worked out sales argument in my hand. Never in my life have I been so humiliated.

The worst of it was there was nothing that I could do but stand around and wait, and I have been waiting a couple of hours. I watched Sam as he took the north fork and went plowing along across the face of the mountain. And I watched the Mammoth take the south fork and go zigzagging up through the woods and finally disappear around the shoulder into the little side valley. I regret to state that the snow is not deep enough to stop the Mammoth tractor.

Of course we are putting on a better demonstration; Sam is making three miles an hour, while the clumsy Mammoth is making hardly more than a mile and a half. But I doubt if that will do us any good. I must be getting old, I am getting pessimistic. But what sense is there, I ask you, in wearing yourself out on an old bozo like Job Chipman, who is deaf and half blind and—I am beginning to think—hopelessly feebleminded?

However, while there is even a spark of hope, Alexander Botts is not the man to lay down on any job. I have come back to the hotel, and in between periods of watching the distant tractors on the mountainside I have been improving my time in writing this report. And in addition, as a last desperate hope, I have prepared an abridged one-page edition of my sales talk. The old idiot ought to be able to read at least that much. It is a mere pitiful skeleton of the splendid thirty-one-and-a-half-page sales argument, but at least it is something.

When Mr. Chipman gets back I will hand it to him, and if he will read it he may get a faint idea of what he will be missing in case he fails to buy the Earthworm.

It is now ten o'clock, and I can see from my window that Sam has finished plowing out his road and is on his way back. Accordingly I will go down to the sawmill and talk things over with him when he arrives.

Later. 2 p.m.

Since writing the above words a most unusual and extraordinary series of events has taken place. But before relating these happenings I wish to remind you of a remark I made in a previous report to the effect that I considered Mr. Sam Simpson one of the best mechanics in the employ of the company. It now appears that I was right, as usual. As a judge of men I am practically perfect. For Sam is not only one of the best mechanics in the country but he is also not so terribly bad as a salesman.

When I left the hotel, after writing the first part of this report, I walked down to the sawmill to wait for Sam. But I was somewhat surprised to see that when Sam reached the fork in the road he did not return to the mill. Instead, he swung up the south fork, following the path which had been taken by the Mammoth tractor. As the road had already been plowed, he was able to go along at full speed and before long be had disappeared—as the Mammoth tractor had done before him—around the shoulder of the mountain. This procedure on Sam's part puzzled me somewhat, but as I have every confidence in his judgment, I assumed that he had some reason for this action. And it turned out that this was the case.

For two hours nothing happened. And then, a little after noon, the Earthworm appeared once more, coming back down the south fork road. It rolled merrily along, and before many minutes it arrived at the sawmill.

The door of the cab opened and out stepped Sam and Mr. Chipman. Mr. Chipman walked right over to me.

"Let's go down to your room at the hotel," he said. "I want to talk to you."

As we walked along I noticed that Sam's honest face was lit up by a smile of perfect peace and happiness.

"It worked!" he said. "It worked!"

"What worked?" I asked.

"A little plan I had," said Sam. "I didn't tell you about it before because I wasn't sure it would succeed, and I didn't want to raise your hopes too high."

"What happened?" I asked.

"After I finished putting on the grousers yesterday afternoon," Sam explained, "I got to talking with one of the mill hands about those roads. He told me that Mr. Chipman would probably have us plow them out. He said the north road is about four miles long and ends in a nice flat open space where it is easy to turn around. But the south road is twenty miles long and you have to go clear to the end before you find any real good place to turn. So that was why I hurried off so fast this morning. I wanted to get to the fork first and take the north road that had the good turning around place."

"You didn't need to do that," I said. "The Earthworm can swing right around anywhere in its own tracks."

"Exactly so," said Sam, "but the Mammoth can't, and I had a feeling that if I could once get old Jones started up that south fork he would be a blowed-up sucker. And I was right."

"Go on," I said.

"After I had finished plowing my road," said Sam, "I chased up the south fork and caught up with the Mammoth about five miles out. I found Mr. Chipman hollering and yelling that it was time to turn around and go home—that he had to get back so he could make an early start on a trip he was going to take down the valley, and that it was a fine tractor that could go only in one direction. And all the time Mr. Jones was writing on the silly little slate, trying to smooth him down by explaining that the road was too narrow and that they would turn around just as soon as they got to a good place."

"So what did you do then?"

"Nothing much," said Sam, "except give them a little demonstration. I ran up and down the road four or five times and let them see how easy

I turned around at the end of each little trip. As soon as Mr. Jones saw how nice the Earthworm did it, he took a terrible chance and tried to turn around with the Mammoth."

"Did he make it?"

"No," said Sam, "he did not. He picked a place where they had made the road extra wide to let the log sleds pass, and he started to swing around very slow and cautious, but all he did was get his big clumsy machine stuck deep down in the ditch beside the road."

"Is he still stuck?" I asked.

"No," said Sam. "We hooked onto him with the Earthworm and yanked him out—all of which gave Mr. Chipman a swell demonstration of how well the Earthworm can pull. So the Mammoth is not stuck any more, but it is still headed in the wrong direction, and it still has about fifteen miles to go before it can turnaround. After I had pulled the Mammoth out, I took Mr. Chipman aboard and brought him back here."

By this time we had reached my room in the hotel. Looking out the window, we could see a little speck moving slowly along far away on the side of South Mountain. It was the great Mammoth tractor, with a long trip still ahead of it. But Sam said it had plenty of gasoline and ought to get back sometime tonight.

As we turned away from the window I decided that the time was ripe for action. I handed out my little one-page abridged sales talk. But I was doomed to one last bitter disappointment. Mr. Chipman shoved it to one side, saying he was not interested in such things.

He went on to say, however, that he had seen such a good demonstration that he wanted to keep the Earthworm and order three more just like it. So, with deep humility—realizing that all my frenzied activity of the past two days had made absolutely no impression on anybody, and that the success of the sale was due entirely to Sam—I signed up Mr. Chipman for four ten-ton Earthworms.

Sam and I are leaving on the afternoon train. And in a very humble spirit, I am writing a polite little note to be given to Mr. Jones when he gets back to the hotel. This note ought to cheer him up a great deal, because it informs him that he is a wise egg and that he was absolutely right when he said that if a man can put on a real good demonstration he doesn't have to have any sales talk.

Very sincerely,
ALEXANDER BOTTS,
Salesman.

SECONDHAND STUFF

ILLUSTRATED BY TONY SARG

FARMERS' FRIEND TRACTOR COMPANY
MAKERS OF EARTHWORM TRACTORS
EARTHWORM CITY, ILLINOIS

JUNE 2, 1922.

MR. ALEXANDER BOTTS,
HOTEL VAN VOORTEN,
ALBANY, NEW YORK.

DEAR MR. BOTTS: At your earliest convenience we want you to call on Mrs. Hannah Watkins, of Bridport, Vermont. Mrs. Watkins has written us that she is the owner of a four-hundred-acre farm and that she is considering the purchase of a ten-ton tractor. We have sent her a letter giving her full information in regard to this machine, and telling her that you would call on her at an early date.

We trust that you will have no trouble in securing her order.

Very sincerely,
GILBERT HENDERSON,
Sales Manager.

FARMERS' FRIEND TRACTOR COMPANY
SALESMAN'S DAILY REPORT

DATE: MONDAY, JUNE 5, 1922, 9 P.M.
WRITTEN FROM: ADDISON HOUSE, MIDDLEBURY, VERMONT.
WRITTEN BY: ALEXANDER BOTTS, SALESMAN.

I got your letter yesterday. I took the train from Albany early this morning, and arrived at Middlebury—the nearest station to Bridport—about noon. After a good lunch I hired a neat roadster from the local undertaker and garage owner, drove out about ten miles to Bridport, and called on Mrs. Hannah Watkins.

This lady is a tough prospect. And that is not all. She has been made a whole lot tougher—almost too tough to handle—by certain so-called information which she has received from the company.

Now I realize that I am nothing but a salesman, and I do not presume to criticize Mr. Henderson, the sales manager. But if I were criticizing him I would point out that when he is sending out a man as good as I am, he had better leave everything to me and not hamper my activities by writing letters giving information to the prospect. And this suggestion applies with double force to the very full information which Mr. Henderson unfortunately saw fit to send in his letter to Mrs. Hannah Watkins. In order that you may see exactly what I mean, I will give a full description of my visit to this lady.

As I drove in the front gate of the farm I was at once struck by the fine appearance of the place. The fields looked rich and beautiful, and the farm buildings were large and in splendid repair. There was such an air of prosperity everywhere that I knew at once the proprietor would have plenty of money to buy a tractor. Stopping my car at the barn, I conversed for a minute or two with one of the hired men, and learned that Mrs. Watkins' husband had died some two years before, that since that time she had been managing the farm herself, and that she was reputed to be very wealthy.

I then walked up to the kitchen door, knocked, and was admitted by Mrs. Hannah herself. She was a small thin woman of about fifty, with what appeared to me to be a very shrewd looking face. I realized later that "shrewd" was a very mild word.

I started my call by getting out my folders and pictures and giving a splendid sales talk. I first described the tractor and its advantages. Then, after asking about the various kinds of work on the farm, I explained in my usual clear and convincing manner how all this work could be done much more cheaply and efficiently by a ten-ton Earthworm than by horses or any other known method.

Mrs. Watkins agreed with me on all of my points, and I had every reason to suppose that I would soon have her name on the dotted line. But such was not the case. I had no sooner completed my explanations than Mrs. Watkins spoke up herself.

"When I wrote to your company a couple of weeks ago," she said, "I asked particularly about the durability of your tractors. I wanted to know whether they would be worth anything after three or four years' use. In answer to my questions Mr. Henderson, your sales manager, wrote me this letter."

She then handed out Mr. Henderson's most unfortunate effort. I have no copy of it, and I cannot repeat the whole thing, but certain sentences stand out in my mind with distressing distinctness. Near the beginning

Mr. Henderson said, "You need have no doubt about the durability of our machines. We made the tractors used by the United States government to pull artillery during the World War. What is good enough for Uncle Sam is good enough for anyone." At another place he said, "Before the war, during the war, and after the war, Earthworm tractors have always been made with the same scrupulous care. They are so well constructed that they will last for many years—even when abused, overloaded, and given insufficient care." And near the end of his letter Mr. Henderson remarked, "An Earthworm three or four years old is only at the beginning of its usefulness; it is worth practically as much as a new machine."

No doubt Mr. Henderson, in writing this letter, supposed that he was producing something very clever—and far be it from me to criticize my boss. I will admit that his arguments, when used on some people, would be great stuff. But on other people they are not so good. And duty compels me to report the effect of his words on this particular prospect.

"What do you think of it?" asked Mrs. Watkins.

As she spoke she regarded me with such a cold, calculating eye that I began to suspect there was going to be some funny business connected with this deal. I hated to commit myself, but I could not very well disagree with the written statements of my own sales manager.

"Mr. Henderson has written you the exact truth," I said.

"Good!" said Mrs. Watkins. "I have an old Earthworm tractor out in the barn, and it's fine to know that it's worth so much. I want to trade it in to you people on a new one."

At this point I knew for sure that there was going to be some funny business on this deal. In fact, I began to suspect that old Hannah was getting ready to try and play me for a sucker.

But I concealed my feelings and merely asked to see the machine which she wished to turn in.

She led me out to the barnyard, where my startled eyes fell upon one of the saddest wrecks I have ever seen in all my experience in the tractor business.

"It's an Army ten-ton," said Mrs. Watkins. "The government turned it over to the highway commissioners after the war, and the highway commissioners sold it to my husband. How much is it worth on a trade?"

I looked it over. It was an Army ten-ton all right, or, rather, it was the decayed and rusted remains of what had once been an Army ten-ton. Apparently, it had been completely worn out by hard work and abuse. Every bolt on the whole machine seemed to be loose. The tracks were

She led me out to the barnyard, where my startled eyes fell upon one of the saddest wrecks I have ever seen in all my experience in the tractor business.

wobbly, the sprockets were out of line; the fan belt had been chewed by rats; and mice had made nests in the seat cushions. There was plenty of play in the main bearings, but no compression in the cylinders.

The armor had been replaced by a crazy, homemade tin hood, but the only other repair work ever done on it had apparently consisted in tying together broken parts with hay wire. I have seen a good deal of hay wire in my life, but never have I seen any one machine that had acquired so much of it. There was also a good deal of binder twine, friction tape and bits of old harness in evidence. And the chickens had been roosting on the old thing all the spring.

It was absurd for anybody to expect to get anything for this machine other than its value as scrap iron. But I decided that we could afford to lose something, provided we could get this lady's order for a new ten-ton, so I promptly offered five hundred dollars. If Mrs. Watkins had had any sense she would have leaped at the offer. But she did not.

"Really, Mr. Botts," she said, "I have no time to listen to your jokes, amusing as they are. A new ten-ton tractor, according to your price list, sells for six thousand dollars. Naturally I do not expect to get that much for a secondhand machine. But I do feel that I ought to have at least five thousand."

"Impossible!" I said. "This machine is three or four years old."

"Exactly," said Mrs. Watkins. She then opened up Mr. Henderson's letter—which she had had the bad taste to bring out to the yard with her—and read me a short extract: "'An Earthworm three or four years old is only at the beginning of its usefulness; it is worth practically as much as a new machine.'"

"But this is nothing but a war tractor," I said. "Our present machine is so vastly improved that there is no comparison between them."

Mrs. Watkins again referred to the letter: "'What is good enough for Uncle Sam is good enough for anybody.'"

"But during the war," I said, "construction was necessarily very hurried. Practically all factories were turning out material so fast that they had no time to pay any attention to quality."

"'Before the war, during the war, and after the war,'" quoted Mrs. Watkins, "'Earthworm tractors have always been made with the same scrupulous care.'"

"Well," I said, "probably this machine was pretty good when it was new. But look at it. Look at the shape it's in. It has been abused. It hasn't had sufficient care. It's a wreck."

"'They are so well constructed,'" quoted Mrs. Watkins, "'that they will last for years—even when abused, overloaded, and given insufficient care.'"

"I am afraid," I said, "that you are taking that letter too literally."

"You said yourself that Mr. Henderson had written the exact truth."

"In a general sense, yes," I replied. "But in this particular case, no."

"Your reasoning," she said, "is absolutely no good. If you back down this way, how can I believe what you tell me about the new machine I am thinking of buying?"

"I can assure you, madam," I said with deep earnestness, "that our new machine will live up to everything I have said about it."

"The only way that you can prove to me that you are dependable," replied Mrs. Watkins, "is to back up the written words of your sales manager by giving me a proper price for my old tractor. I'll tell you what I'll do: I'll knock off a thousand dollars. But four thousand is my rock-bottom price."

"You'll never get it."

"Maybe not," said Mrs. Watkins. "But in that case I'll do business with a company I can trust. The salesman for the Steel Elephant Tractor Company is coming around next week, and I'll trade in my machine for what I can get on a new Steel Elephant."

"That would be a mistake," I said.

"I have made up my mind," said Mrs. Watkins. "If I can't get four thousand out of you people I won't deal with you."

"But four thousand is impossible."

"If you change your mind within the next few days," said Mrs. Watkins, "you will find me at my daughter's house in Middlebury. I am going there tonight for a week's visit. Good afternoon." And she turned around and walked into the house.

I was left, I must admit, in something of a quandary. My first impulse was to pursue this lady into the house and resume the argument. But what chance would I have? Every time I opened my mouth Mr. Gilbert Henderson—represented by that unhappy letter—would rise up and call me a liar. Ordinarily I could have handled Mrs. Watkins very easily, in spite of her stubborn and unreasonable nature. But now that her mind had been poisoned by the insidious propaganda emanating from the home office itself, I felt that I ought to think things over before proceeding any further.

Accordingly I returned to Middlebury, where I have arranged to spend the night at the Addison House. I will sleep on this problem, and perhaps tomorrow I may figure out some different method of approach which will cause this crazy old lady to listen to reason. I am not licked yet.

> Yours,
> ALEXANDER BOTTS.
> *Salesman.*

FARMERS' FRIEND TRACTOR COMPANY
SALESMAN'S DAILY REPORT

DATE: JUNE 6, 1922.
WRITTEN FROM: MIDDLEBURY, VERMONT.
WRITTEN BY: ALEXANDER BOTTS.

As soon as I woke up this morning, I had a brilliant idea. I at once acted upon it and everything would have been fine except for the sudden appearance of another incredible letter written by Mr. Gilbert Henderson. As it is, the situation is in a good deal of a mess.

No one should make the mistake of thinking that I am criticizing Mr. Henderson. My loyalty to the company and to my boss is as strong as ever, but I must admit that something of my old self-confidence and assurance is gone. From now on, every time I approach a prospect it will be with the haunting fear that he may have up his sleeve some weird, outlandish communication from the home office that will knock my plans completely cockeyed. When I relate the events of the day you will see that my fears have justification.

The idea which came to me this morning was a good one. I remembered that some time ago in Albany one of the salesmen of our company had told me of a call he had made on a Mr. George Anthony of Fort Henry, New York, which is just across Lake Champlain from here. It seems that Mr. Anthony had been interested in getting a secondhand tractor. He had been offered an old Army ten-ton which is still in our Albany warehouse, but for some reason had not taken it. If this Mr. Anthony still wanted a secondhand tractor it occurred to me that I might be able to get him to make an offer on Mrs. Watkins' machine. If I could get him to offer a couple of thousand I might possibly talk Mrs. Watkins into taking it. It was worth trying.

Accordingly I rented the same car which I had yesterday, drove down to Lake Champlain, crossed on the ferry, and about the middle of the morning reached Mr. Anthony's farm outside Port Henry.

Mr. Anthony turned out to be a young man of pleasant personality, and we were at once on a very friendly footing. It appeared that we had both been in the artillery during the war and had both been most favorably impressed by the tractors which were used to pull the guns.

"Yes, sir," he said, "I drove one of those artillery ten-tons for several months in France. They are the finest tractors ever made. I've often thought of buying one secondhand to use here on the farm."

"You would rather do that than get one of our new, improved machines?"

"Absolutely. I want exactly the same model I drove on the other side."

"I understand one of our salesmen offered you one last year."

"Yes, he had one at Albany he wanted to sell me for five hundred dollars."

"What was the matter?" I asked. "Was the price too high?"

"No," he said, "the price was too low. I didn't even go down to look at it. I knew that if he was offering it as cheap, there must be something the matter with it. I want a machine that is in fairly food shape."

At these words I began to feel that my visit was in vain. If this bozo wanted an expensive machine I could accommodate him fine, but I wasn't so sure about the "fairly good shape" business. However, I decided not to give up without a struggle.

"I have exactly the machine for you, Mr. Anthony," I said. "It is across the lake. If you will get in my car I will drive you over to look at it."

"Fine," said Mr. Anthony, "I would like nothing better."

After telling a couple of hired men what to do while he was gone, and saying goodbye to Mrs. Anthony in the house, he climbed into my car, and we drove down to the ferry. After crossing to the Vermont side and getting a few sandwiches at a hot dog stand, we finally reached the Watkins farm shortly after noon.

One of Mrs. Watkins' hired men took us out to the barnyard, and as he showed us the machine I will admit that my heart sank within me. In my enthusiasm over the possibility of making a sale I had forgotten what a truly horrible looking mess of junk this tractor was. For a moment I was speechless, but Mr. Anthony at once began to talk.

"Yes, sir," he said, "this is the genuine article—exactly the same kind of bus that I used to drive in the Army. I used to hate the Army, but now I look back and I know that them were the days. And this is the finest model tractor that was ever built."

"It might have been worth something once," said the hired man, "but it's pretty old and rusty and dirty now."

"All that rust and dirt don't amount to anything," said Mr. Anthony. "What counts is the machinery inside."

"Most of that machinery is just throwed together," the hired man went on. "The road commissioners we got it from said it was sort of built up out of the best parts from a couple of old wrecks of machines as the government gave them."

All this time I was trying to signal the hired man to keep his mouth shut, but he was too dumb to understand. Fortunately, Mr. Anthony was so busy looking over the machine that he didn't pay much attention.

"This certainly takes me back to the old wartimes," he remarked. "It's just like meeting an old friend. I haven't seen one of these machines since I got my discharge."

"The compression is awful weak," said the hired man helpfully.

"As soon as I put in new rings and grind the valves," said Mr. Anthony, "she'll be as good as new. I can hardly wait to get started overhauling the old baby. These motors are so accessible and handy it's a pleasure to work on them."

"A couple of them radiator sections leak pretty bad," said the hired man.

"That's all right," said Mr. Anthony. "Those sections are removable. I'll take them out and solder them in no time at all. Can we start up the motor and see how she sounds?"

"We can try," said the hired man. He primed the cylinders and gave the crank a few flips. Nothing happened.

"Here," said Mr. Anthony, "you don't know how to handle this thing. Give a chance to an old-timer that knows his business."

He climbed up onto the tractor. And while he was priming the cylinders again and adjusting the spark and throttle levers, I got the hired man off to one side. There was a nice pick handle in the corner, but although it would have been a public service to beat in the top of this yokel's empty head, I decided to use more conservative methods. I gave him a dollar on condition that he would go out behind the pigpen and stay there. He went.

I have repeated all of the hired man's conversation, because the dollar I gave him is entered on my expense account, and I wanted to make it clear that this was a necessary expense.

When I got back to Mr. Anthony he was spinning the crank with great energy, but no results.

"She's all right," he said, "only she hasn't been used for a long time, and of course she's hard to start."

He rested a minute, then took out the spark plugs, filed and set the points, squirted oil in the cylinders, put back the plugs, filed the breaker points on the magneto, cleaned the distributor brushes and finally spun the crank again. This time the machine gave a feeble bark, and after a few more spins of the crank it started up with a roar. It hit on only three cylinders, it poured out clouds of blue smoke, and it rattled and clanked and knocked in a manner that was fearful to listen to.

But it ran. Mr. Anthony grinned happily, sat down in the seat and drove around the barnyard. The transmission gears howled, the tracks flopped loosely over the sprocket, and the whole machine shook and vibrated as if it was going to fall apart. After a short drive Mr. Anthony brought it back and shut off the motor.

"When I get these transmission gears adjusted, and the tracks tightened up, and the motor overhauled," he said, "she'll be practically as good as new. How much did you say they wanted for this machine?"

"Four thousand dollars," I said, speaking in a casual, offhand way.

"I suppose," said Mr. Anthony, "that the terms will be the same cash-on-delivery proposition which was described in the letter which your company wrote me when they offered me that other secondhand machine?"

"That is exactly what I had in mind," I replied.

"Sold!" he exclaimed. "Will it be all right if I take it home this afternoon?"

"I'll have to see the owner, who is in Middlebury."

"Do I get any discount for cash?" he asked. "I have the money right in the bank."

"The terms are cash," I said, "and the price is so low that we can't afford to allow any discount."

"Well, that will be all right," he said. "And by the way, if you want any references regarding my standing you can inquire at the bank in Port Henry."

"Fine," I said. "I will drive into Middlebury right now and close the deal with Mrs. Watkins."

"And I will stay here," said Mr. Anthony, "and work on the machine so as to be sure it will be all right to drive over to my place."

As fast as I could I climbed into my rented roadster and started for Middlebury. On the way, however, it suddenly occurred to me that I had better check up a little on Mr. Anthony before I got too deep in this business. Mr. Anthony's willingness to pay four thousand dollars for this bunch of junk might be explained by his memories of the good old days of

1918, and by his sentimental but perfectly natural desire to have a machine exactly like the one he had driven in those stirring times.

But, on the other hand, it might be something else. I have never forgotten the important looking gentleman who stepped into the main office at Earthworm City several years ago, explained that he was the owner of a large ranch in Texas, and electrified the whole sales force by signing up an order for twenty-five ten-ton tractors and nonchalantly making out a check for one hundred and fifty thousand dollars. I have never forgotten this exciting incident, nor have I forgotten the subsequent arrival of the keepers who took the unfortunate gentleman back to the asylum.

Consequently, as I am always cautious and never take any chances of any kind, I turned my car around, drove down to the ferry, crossed to Port Henry, and stopped in at the bank which Mr. Anthony had given as a reference.

The banker assured me that I need have no fears about Mr. Anthony. The farm was only a small part of the property which he had inherited from his father, and he had ample means to purchase several dozen tractors if he so desired. Furthermore it appeared that Mr. Anthony was very highly regarded in the community, being a public apprised citizen and completely honorable in his business dealings.

Thus reassured, I came back on the ferry, drove to Middlebury, and called on Mrs. Watkins. I told her I had decided to allow her three thousand dollars for her old tractor.

As first she held out for four thousand, but we finally compromised on thirty-five hundred, and she signed an order for a new ten-ton, giving me a bill of sale for the old tractor, and agreeing to pay the balance of twenty-five hundred on the delivery of the new machine.

Thus by skillful bargaining I had arranged matters so that we would make our usual profit on the sale of the new ten-ton, and would rake in an additional five hundred dollars on the secondhand proposition. Everything would have been fine had it not been for the unfortunate way in which the home office persists in writing letters of information.

As soon as I had completed my business with Mrs. Watkins, I drove back to the farm and told Mr. Anthony that the deal was closed and that I would be glad to have his check for four thousand dollars at once, so that I could mail it in to the company.

"I haven't got my check book with me," said Mr. Anthony. "And besides, you said that the terms would be the same cash-on-delivery proposition which was outlined in the letter which your company sent me regarding that other secondhand tractor."

"Exactly," I said. "Cash on delivery."

"Here," said Mr. Anthony, "is the letter." He drew from his pocket a communication which he had received last year from Mr. Gilbert Henderson. As I have pointed out before, I have no desire to find fault with the actions of high officials of the company. Therefore I will make no comment on this letter. All I will do is repeat one paragraph so you can see how the best efforts of the finest salesman may be completely jolted out of gear by the ill-advised writing of letters from Earthworm City.

"Our terms," said Mr. Henderson's letter, "will be cash on delivery. Delivery will be considered complete and payment due when you have used said machine for three days and are convinced that it is satisfactory in every way. At the end of three days, if you are not satisfied, you may return the machine at our expense."

You can well imagine my feelings on reading these words. I had always imagined that cash on delivery meant cash right away. And here Mr. Henderson—in order to give unusually attractive terms on that tractor at Albany which nobody seemed to want—had defined cash on delivery in such a way that it meant cash after three days, provided the customer was entirely satisfied and happened to feel like giving it to us. As I contemplated the dismal junk pile I had just bought for thirty-five hundred dollars I began to feel distinctly sick at my stomach. How could anybody be satisfied with the wretched thing after three days to think it over?

Mr. Anthony's next remarks were not reassuring. "I've got several hours of daylight yet," he said. "I'll take it over on the ferry this afternoon, and Friday night—if everything goes well—I'll send you my check."

At once I began figuring up some sort of an argument to get the money right away. But as I have stated before, it is a difficult thing to argue directly against the written statements of your own sales manager. And while I was thinking as hard as I could Mr. Anthony cranked up the motor and prepared to drive away.

"Don't you want me to help you?" I asked.

"No," he said. "I can manage all right myself."

"Then I guess I'll go back to Middlebury for the night."

"Will I see you again?"

"You bet you will," I said. "I'll be over tomorrow to find out how you are getting along."

"All right," he said. "See you tomorrow." And he drove the tractor, clanking and roaring, down the road in the direction of the ferry.

In a very worried frame of mind I returned to Middlebury, where I have been spending the evening writing this report. Tomorrow morning I will go over to Mr. Anthony's farm, hoping and praying that the machine will hold together that long, and that I may be able to get this money out of him.

In closing I am tempted to make a few remarks regarding the writing of letters, but my loyalty to the company restrains me.

Yours,
ALEXANDER BOTTS,
Salesman.

FARMERS' FRIEND TRACTOR COMPANY
SALESMAN'S DAILY REPORT

DATE: JUNE 7, 1922.
WRITTEN FROM: MIDDLEBURY, VERMONT.
WRITTEN BY: ALEXANDER BOTTS, SALESMAN.

It gives me great pain to report that things are even more complicated than yesterday. And although the present unfortunate situation is in no way my fault, I cannot help feeling greatly depressed. When I tell you the latest developments you will see just why I feel as I do.

Bright and early this morning, I rented the same car, and drove down to the lake. I had no sooner run the car onto the ferry than I was surprised and startled to observe the old ten-ton tractor standing up near the bow.

There were a number of automobiles on board, and a couple of dozen people, but no sign of Mr. Anthony. As soon as we got under way I engaged one of the crew in conversation.

"What is this tractor doing here?" I asked.

"I'll tell you one thing it's doing," said the member of the crew. "It's cluttering up half our deck space and getting in the way of all the automobiles that goes on and off, and making us more trouble than a dead whale."

"How did it get here?"

"The poor boob that owns it drove it on last night. Where he made his big mistake was when he shut off the motor."

"Couldn't he get it started again?"

"He could not. It wouldn't even cough. He worked on it for an hour, while we was making a couple of trips back and forth. Finally he said he thought the magneto had gone dead on him, so he took it off and carried it away with him to get it fixed, and we ain't seen him since."

"That was certainly tough luck," I said.

"If you ask me," the man went on, "I would say it ain't only the magneto. I would say the whole blooming machine has gone dead on him. Did you ever in all your life see such a sad looking bunch of stove iron?"

"No," I said, "I doubt if I ever did."

"Well," continued the man, "he'd better get it off here pretty soon or his ferry bill will be more than his old buggy is worth."

"You don't mean to say that you are charging him for each trip you make?"

"I'll say we are. Traffic is heavy right now and this thing gets in our way something terrible."

"Couldn't you tow it off with a truck or something?"

"Not a chance. We might be able to pull it off the boat, but it's so heavy we'd never get it up the narrow, steep landing. And if it got stuck on the landing it would block the whole passage. The guy should of had more sense than to try to drive a machine that has gone completely hay wire like this. He said he bought it secondhand and I guess he's sorry now he ever got it."

"Very likely he is," I said sadly.

When the boat got to the other shore I decided that this was not an auspicious moment to strike Mr. Anthony for payment. There are times when silence is golden, and when even the best of salesmen is better off at a distance from the purchaser. With these thoughts in mind, I left my car on the boat, voyaged back to the Vermont side, and returned to Middlebury.

I at once called on Mrs. Watkins to find out what her attitude would be in case we wished to call off the deal and return her secondhand tractor. Her attitude, unfortunately, was very unreasonable.

"Most certainly I will not take back that machine," she said. "You have signed the papers. I have given you a bill of sale. You have removed the property. If you try to back out I will sue you."

As I could get no satisfaction from this most disagreeable woman, I returned sadly to the hotel, where I cheered myself up slightly by eating a very large and excellently cooked luncheon.

Late in the afternoon I drove down to the ferry once more, and when the boat came in I was shocked to see that the tractor was still on board. I did not even ask whether Mr. Anthony had come back. Feeling that the time was not yet ripe for an interview, I returned to Middlebury, meditating sorrowfully on Mr. Henderson's curious definition of cash on delivery, and on his startling estimate regarding the value of a tractor three or four years old.

Tomorrow morning I will make another accounting expedition, and if Mr. Anthony has at last succeeded in getting his machine running I will attempt to approach him and see what can be done. For the present I can do nothing but sit around sadly, in my hotel room, while my brain is tortured by the sickening thought of that tractor riding back and forth, back and forth, on that ferryboat, reeling up a stupendous ferry bill.

Yours,
ALEXANDER BOTTS,
Salesman.

FARMERS' FRIEND TRACTOR COMPANY
SALESMAN'S DAILY REPORT

DATE: JUNE 8, 1922.
WRITTEN FROM: MIDDLEBURY, VERMONT.
WRITTEN BY: ALEXANDER BOTTS, SALESMAN.

Things have been moving today with starling rapidity. This morning I drove down to the ferry. The tractor was gone and the member of the crew whom I had talked with yesterday said, "Yes, the guy that owned it came back with the magneto last night about six o'clock. It seems it had taken all day to get the magneto repaired. And when he put it on, the tractor started up fairly well."

"And how did he act when you asked him for ferry charges for each trip?" I asked.

"Well, he seemed pretty sore," said the man, "and we jawed around back and forth for quite a while. But he finally paid up and went on his way."

I was not much reassured by hearing that Mr. Anthony "had seemed pretty sore," but I decided to go on and make cautious reconnaissance.

As I drove up the lane which led from the main road to Mr. Anthony's farmhouse I suddenly discovered the tractor at the side of the lane in the bottom of a shallow dry ditch. A small boy was standing beside it, looking at it.

"Hello, sonny," I said. "Do you know Mr. Anthony? Is he around here anywhere?"

"Mr. Anthony is my father," replied the small boy, "and he'll be back in just a minute. He's gone up to the barn for some tools."

At this point I suddenly observed that the motor of the tractor was no longer in its normal place under the hood. It had been removed from the machine and was lying on the ground beside it.

"What is the idea?" I asked. "Why did your father take out the motor?"

"He didn't take it out," answered the boy. "It jumped out."

"Jumped out!" I said. "What do you mean? How could it jump out?"

"Well," explained the boy, "it was like this: Father was afraid to go over that little plank bridge with such a heavy tractor, so he drove it down off the road, straight across this ditch. I guess he forgot the ditch was so deep, so he came pretty fast. The machine got a terrible jounce and the motor just jumped out."

"That's impossible," I said. "I never heard of such a thing."

"Neither had father," said the small boy. "He certainly was surprised. But after he looked things over he told me how it happened."

"What did he say?"

"He said the motor was supposed to be bolted on to the frame by four big bolts. But these bolts had all worked loose and dropped out, and the people that owned the tractor had tried to fasten the motor in with a lot of hay wire. It looks like the hay wire wasn't strong enough."

I got out of my car. I walked down into the ditch. I looked over that tractor. And it was all true; the motor had indeed jumped out. It was lying on its side in the grass. The radiator hose connections, the gas line and the control rods had been yanked out by the roots. The clutch had come apart at the coupling, strewing disks in all directions. And although the crank case, cylinders and flywheel had not been hurt, the push rods and rocker arms and manifolds and various other gadgets were in a sad mess.

For half a minute I gazed at this wreck in awed astonishment, and then started back for my car. I had a strong hunch that the proper time for a visit with Mr. Anthony had not yet arrived. But just as I was climbing into the seat Mr. Anthony appeared from the direction of the house and barns. He recognized me at once.

"Hello, there, Mr. Botts!" he yelled. "Wait a minute! I want to talk to you!"

With fear in my heart I got out and waited for him to approach. He walked up to me and stuck out his hand I ducked. I thought he was going to hit me. But he only wanted to shake hands. And after we had shaken hands he started to talk.

"I've been having a little trouble," he said.

"So I see," I remarked cautiously.

"Had a little grief on the ferryboat coming over," he went on, "but it didn't amount to anything—just the magneto. This is more serious, but I can get her fixed up all right."

"Of course," I said. "Stick the old motor back. Perfectly simple, straightforward job."

"It was all my own fault," he continued. "If I had looked this baby over as carefully as I should have I would have seen that the motor was only fastened with hay wire and I would have saved myself considerable trouble. But there's no use crying over spilled milk."

"That's right," I agreed. "Never cry over spilled milk."

"And aside from these few little troubles, she certainly is a swell machine, and I want to thank you for giving me the opportunity to get her."

"You're entirely welcome," I said.

"It gave me a real thrill," he went on, "to come driving along up here last night. It took me back to the good old days in France when I was in the Army. The motor sounded wonderful. It was hitting on only three cylinders, but those three were hitting great. And by the way, I won't be able to try it out very much the next few days, so I might as well pay you now."

He took out his check book and began to write, while I stood there in a trance. And then all of a sudden it came over me what had happened to this guy. By thinking about the good old wartimes he had hypnotized himself to the point where he felt he absolutely had to have this machine. He had provided his own ballyhoo, and he hadn't been pushed into buying by any outside high-powered salesmanship. Having done all the selling himself, he couldn't get sore at the tractor—no matter what happened—without admitting that his own judgment had been all wrong.

And that isn't all. When he handed me the check it was for forty-five hundred dollars instead of the four thousand we had agreed on. The check was good too. I had it cashed later in the afternoon, just as a precaution in case Mr. Anthony should change his mind and try to stop payment; and I enclose the full amount as a draft payable to the company.

Mr. Anthony's explanation of the extra five hundred dollars was most interesting.

"Do you remember," he asked me, "that hired man over at Mrs. Watkins' place?"

"Could I ever forget him?" I said. "And I am glad you paid no attention to what he said."

"But I did pay attention," said Mr. Anthony. "I've been thinking over what he told me about using one old tractor to furnish parts for repairing another one. I will need a lot of parts to fix up this machine, and I have decided that the cheapest way to get them would be for me to buy that five hundred dollar ten-ton at Albany. To save the trouble of writing two checks, I have made this check cover the price of both machines. That will be all right, won't it?"

"Certainly, sir," I said. "And I will have the Albany ten-ton shipped up at once. It is a splendid idea."

"Yes," said Mr. Anthony, "for only forty-five hundred dollars, I get this machine and enough parts to fix it up so that it will be far better than any new tractor at six thousand dollars."

"Isn't it wonderful?" I said. And a few minutes later, when I started for Middlebury on my way back to Albany, I left Mr. Anthony with a smile of complete happiness and satisfaction spread all over his face.

Yours,
ALEXANDER BOTTS,
Salesman.

THE BIG ENDURANCE TEST

ILLUSTRATED BY TONY SARG

THE BIG ENDURANCE TEST

Farmers' Friend Tractor Company
Makers of Earthworm Tractors
Earthworm City, Illinois

September 5, 1922.

Mr. Alexander Botts,
Hotel McAlpin,
New York City, New York.

DEAR MR. BOTTS: We want you to take charge of our exhibition at the Marblebury, Vermont, County Fair, which takes place Thursday, September 14, to Saturday, September 16, inclusive. We have already shipped to Marblebury the ten-ton tractor, tent, literature and various supplies which we have used in recent fairs in Western New York. Our service man, Mr. Samuel Simpson, has been ordered to be on hand to take care of the mechanical work.

You had better go up several days ahead of time to make all necessary arrangements. You will be in complete charge, and it will be your duty to see that our exhibit is made as attractive as possible, to talk to possible customers, to stimulate interest in the Earthworm tractor and to make any sales that you can.

Very sincerely,
Gilbert Henderson,
Sales Manager.

Farmers' Friend Tractors Company
Salesman's Daily Report

Date: September 9, 1922.
Written from: Marblebury, Vermont.
Written by: Alexander Botts, Salesman.

I got your letter day before yesterday. I arrived here yesterday morning. And after two days' hard work I am pleased to advise that I am getting ready to put on what will be the most stupendous exhibition ever staged by the Farmers' Friend Tractor Company, or any other company—and this in spite of the fact that I am entirely new at county fair work.

When you gave me this job you no doubt expected that I would handle it in my usual competent manner, but when I tell you what I have been doing here you will be amazed. You will see that I have gone far beyond anything that you could have asked or hoped. Besides our regular exhibit, I am arranging added sensational features which will make the eyes of the town folk pop out of their heads. I will give them action, suspense and the highest kind of authentic appeal.

In the first place, I have arranged a thrilling endurance test, and I have arranged it so cleverly that it will cost the company nothing. When I first got here I found that there is a man by the name of Eben Lockwood living about a mile from town who owns one of our five-ton Earthworm tractors. I at once called on Eben and was pleased to discover that he is an enthusiastic booster for the Earthworm.

"Yes, sir," he said, "that machine of mine is two years old, and it's better now than the day I bought it. It has just been overhauled. Tomorrow I am going to start two hundred and fifty acres of fall plowing, and in about two weeks I'll be through. When I plow with that machine I do the work quicker and easier than any other farmer in the state."

As I listened to these words my subconscious mind all at once gave birth to a brilliant idea.

"Mr. Lockwood," I said, "you ought to go after the world's nonstop tractor endurance record."

"What's that?" he asked.

"Last year," I said, "a tractor in Minnesota, driven by two operators, working in shifts, ran five days and five nights without stopping. This was a world's record. The owner at once became famous. He made thousands of dollars in vaudeville and is now running for Congress."

Note: As far as I know, there never was a tractor in Minnesota or anywhere else that ever made any such record or caused any such stir in vaudeville or congressional circles. But I had to set up some sort of mark to shoot at.

"I'd like to see you go after that record," I continued. "But probably it wouldn't be much use. You wouldn't have a chance in the world."

"You're crazy," said Mr. Lockwood. "What they can do in Minnesota, we can do in Vermont—and a whole lot better."

"Well," I said, "instead of running in the daytime for two weeks, why don't you run day and night for one week? It won't cost you any more."

"I'll do it!" said Mr. Lockwood. "My son and I will drive in shifts. We'll run seven days and seven nights without a stop."

"Fine!" I said.

And before he had a chance to change his mind I made final arrangements with him. We fixed up some tanks on one of his wagons so we could drive alongside the tractor and run in gasoline, water and oil without having to stop. We installed a battery under the tractor seat and connected it to an old automobile headlight which we mounted on the front of the tractor to give light for night driving. Then I went into town and persuaded the county farm agent to act as starter and timekeeper. This will give our record-breaking performance a very scientific and official aspect.

This afternoon at five o'clock the great run began. It will end at five o'clock on the afternoon of next Saturday, which is the last and biggest day of the Marblebury Fair. I have agreed to give Mr. Lockwood all the publicity I can. And if he succeeds I will send notices to all the papers and put up a large placard at the county fair proclaiming that he is the tractor endurance champion of the world. That is all I have to do. Mr. Lockwood does all the work and pays all the expenses.

I do not know whether Mr. Lockwood intends to go into vaudeville or run for Congress. But I am practically certain that he will be able to keep going for the full seven days and nights. We have run tractors at the factory much longer than a week. And the only reason nobody has ever made a record like this in actual field work is because no one except myself ever had enough creative imagination to think of it.

I am naturally very proud of having conceived this splendid endurance-record idea. But I am even more proud of the fact that I have gone on and evolved other plans of even greater brilliancy. My logical mind at once perceived that this endurance test will not give us any great amount of publicity unless people come out to see it. And in order to attract the crowds I have made up my mind to inaugurate the First Annual Marblebury Bathing Beauty Contest. I have discovered that there has never before been such a contest in this region—possibly due to the fact that there is no water here and that the town folk erroneously suppose that you must have water in order to have bathing girls. Nothing, of course, could be further from the truth.

It is my intention to open this contest to any girl in the whole county who wishes to enter, to appoint a judging committee of leading citizens and to offer prizes of as many hundred dollars as the company will authorize. I will then hire the local Marblebury brass band for an hour on each of the three days of the fair. This can be done at a total cost of only thirty dollars, which is very reasonable when you consider that the band, although perhaps not very good, is tremendously powerful and can make a lot of noise.

The first two afternoons of the fair I will have a parade from the fairgrounds out to the farm where the endurance test is taking place. The brass band will march in front, next will come a float with the bathing beauties, and following this I confidently expect to have a large and representative crowd. The citizens of this town seem to be a splendid lot—interested to an unusual degree in the higher things of life. They will be attracted into the parade because of the educational and artistic appeal of the bathing beauty contest, and they will follow along to Mr. Lockwood's farm, where they cannot help being impressed by the big nonstop endurance test.

At five o'clock in the afternoon of the last day of the fair, when the tractor has completed seven days and seven nights of running, it is my intention to have it come into the fairgrounds and run two or three times around the race track, preceded by the brass band and followed by the bathing beauties. Anyone can see that this exhibition will completely back off the map anything that has heretofore been attempted. It will give us more favorable publicity than any ten ordinary tractor shows put together.

When I first thought of the bathing beauty idea I was all eagerness to proceed with the arrangements at once. But as the contest will cause a certain amount of unusual expense, and as the company in the past has sometimes seen fit to protest certain items in my expense accounts, I have decided to wait until I receive permission from the company. As soon as this report is received, I would like you to telegraph me authorizing the expenditure of thirty dollars for the brass band and five hundred dollars as prizes in the First Annual Marblebury Bathing Beauty Contest. And as soon as I receive your wire I will go ahead and put on one of the greatest and most magnificent tractor exhibitions that has ever been known.

Yours,
ALEXANDER BOTTS,
Salesman.

THE BIG ENDURANCE TEST

FARMERS' FRIEND TRACTOR COMPANY
SALESMAN'S DAILY REPORT

DATE: SEPTEMBER 11, 1922.
WRITTEN FROM: MARBLEBURY, VERMONT.
WRITTEN BY: ALEXANDER BOTTS, SALESMAN.

It gives me great pleasure to report that the endurance test is going swell. The tractor is pulling a four-bottom plow and turning over between thirty and forty acres each twenty-four-hour day. Mr. Lockwood is driving the night shift from five in the evening until five in the morning, and his son is driving from five in the morning until five in the evening. They change shifts without stopping the machine at all, and the arrangements for putting in gas, oil and water are working perfectly.

The tractor is running splendidly and there is no reason why it can't keep on for the rest of the week. The only trouble is that practically nobody is coming out to see this magnificent record-breaking performance. But by the time the fair opens on Thursday I confidently expect that I shall have received a telegram from the company authorizing me to go ahead with my bathing girl program.

As yet I have made no definite arrangements. But in my quiet way I have been talking the matter up with various people in the town. I have interviewed a number of the leading merchants, and they have admitted to me privately that they would like nothing better than a chance to be on the board of judges in a bathing beauty contest. I have also talked with a good many prominent citizens whom I met informally standing in front of the post office or loafing around Baxter's Garage. They all assured me that the bathing beauty contest would be a splendid thing. It was the unanimous opinion that it would be one of the high points in the cultural and aesthetic life of the community.

This afternoon I mentioned the matter to the very good-looking young lady who runs the soda fountain in Hopkins' Drug Store. As soon as she heard that there would probably be five hundred dollars in prizes she said that she would most certainly enter. My plan has also received favorable comment from a number of other pretty girls, including the bookkeeper at Johnson's Hardware Store, the young lady who runs the telegraph office and a number of waitresses at the hotel. We are sure to have a large and very classy entry list. And as soon as I receive your telegram of authorization the excitement will begin.

Yours,
Alexander Botts,
Salesman.

Farmers' Friend Tractor Company
Salesman's Daily Report

Date: September 12, 1922.
Written from: Marblebury Vermont.
Written by: Alexander Botts, Salesman.

 The big tractor Marathon is still running along as steady as clockwork. I have put placards all over town advertising the great event, and I have had a splendid notice in the local paper, but very few people come out to see it. I am still anxiously awaiting your telegram.
 Mr. Samuel Simpson arrived in town this morning, but the exhibition tractor, the tent and other supplies which were shipped from Syracuse, New York, ten days ago have not yet showed up. And although the endurance test and the bathing beauties will be the most important part of our exhibit, we want this other stuff too. The freight agent is sending out telegrams in an attempt to trace this shipment and if possible hurry it along so as to have it here when the fair opens day after tomorrow.
 I stopped in at the fairgrounds today and noticed that the Steel Elephant Tractor Company is going to have a very large and pretentious exhibit. It seems that Mr. George Crossman, who has the local agency for the Steel Elephant Company, is one of the big business men of the town and is president of the Marblebury Fair Association. He has fixed it so that the Steel Elephant people have the most prominent position, right opposite the main gate, and he has given us a very obscure plot of ground away off in one corner of the fence. I saw Mr. Crossman today and protested against this favoritism, but I regret to say that he only laughed at me. He seems like a very disagreeable person and actually presumed to make some

rather sneering remarks about the great endurance test. Furthermore, he made some dirty remarks about my other plans.

"I hear," he said, "that you are planning to put on a bathing beauty contest."

"What if I am?" I said.

"Nothing," said Mr. Crossman, "only this: As a decent and respectable citizen, I am against any such vulgar and degrading exhibition. And as president of the fair association I will refuse you permission to hold it on the fairgrounds."

"All right," I said, "I will hold it outside. And it will not be vulgar or degrading. It will be beautiful, artistic and inspiring. The trouble with you is you are jealous because you didn't think of it yourself. . . .. Good afternoon."

After thus putting him in his place I withdrew with my usual dignity. And as soon as I receive your telegram I will go ahead without paying any attention to the evil-minded old buzzard. If I can't have my bathing beauty contest at the fairgrounds, I will have it at Mr. Lockwood's farm and draw off half the crowd from the fair.

> Yours,
> ALEXANDER BOTTS,
> *Salesman.*

TELEGRAM

MARBLEBURY VERMONT SEPT 13 1922
FARMERS FRIEND TRACTOR COMPANY
EARTHWORM CITY ILLINOIS

MARATHON RUNNING FINE BUT NO WORD FROM YOU ABOUT BATHING GIRLS PLEASE WIRE ME AT ONCE

ALEXANDER BOTTS

NIGHT LETTER
EARTHWORM CITY ILLINOIS SEPT 13 1922
ALEXANDER BOTTS
MARBLEBURY VERMONT

YOUR REPORT AND WIRE RECEIVED WE DO NOT AUTHORIZE ANY EXPENSES FOR BATHING BEAUTY CONTEST OR BRASS BAND FURTHERMORE WE EMPHATICALLY DISAPPROVE SUCH METHODS OF PUBLICITY STOP THIS COMPANY IS IN FARM MACHINERY NOT BURLESQUE SHOW BUSINESS OUR EXHIBIT MUST BE ATTRACTIVE AND INTERESTING BUT ALSO CONSERVATIVE AND DIGNIFIED

FARMERS FRIEND TRACTOR COMPANY
GILBERT HENDERSON
SALES MANAGER

Farmers' Friend Tractor Company
Salesman's Daily Report

Date: September 14, 1922.
Written from: Marblebury, Vermont.
Written by: Alexander Botts, Salesman.

Your night letter arrived this morning, and I will have to admit that I am deeply disappointed. I am a loyal employee and I do not intend to criticize my superiors, but it is my painful duty to report that the refusal of the company to permit me to carry out my publicity plans will make all my work in connection with this Marblebury Fair go for absolutely nothing.

The bathing beauty contest has not only got away from me, but it has gone over to the camp of the enemy. When I announced that it was all off, the girls who had expected to enter got very sore. They held an indignation meeting under the leadership of the young lady who runs the soda fountain in Hopkins' Drug Store. For some unknown reason they made a protest to Mr. George Crossman, and this slimy snake in the grass at once told them that he would put on the contest himself.

As might have been expected, he is going to do it in a very stingy, cheap sort of way, offering a miserable little first prize of only fifty dollars, and no other prizes at all. The girls were much disappointed about this and said they could not afford to waste three whole days for such a wretchedly inadequate prize. It was finally agreed, however, that they would appear on Saturday afternoon only, and eighteen of them promised to enter. Thus you see that this big bum, George Crossman, has not only stolen my splendid idea but he is going to carry it out in such a cheap and sordid manner that it will be a disgrace to the entire town. Instead of uplifting the moral tone of the community, it will merely provide coarse amusement for the lowest elements in the population. And, in addition, it will undoubtedly give a certain amount of low-grade but nevertheless effective publicity to the Steel Elephant Tractor. The situation is indeed pitiful.

This morning the fair opened. Right opposite the entrance is a tremendous tent with many flags flying, and with two large signs out in front. One sign says:

THE STEEL ELEPHANT TRACTOR IS THE BEST OF THEM ALL.

And the other says:

SATURDAY AFTERNOON, 4 P.M., FIRST ANNUAL MARBLEBURY BATHING BEAUTY CONTEST UNDER THE AUSPICES OF THE STEEL ELEPHANT TRACTOR COMPANY ADMISSION FREE

Inside the tent are four different models of the Steel Elephant Tractor, all painted up like new manure spreaders and making a very brave display. There is a long table covered with literature, and there are no less than four attendants, who are spending their entire time pouring their insidious propaganda into the ears of the honest but credulous country folk who have come in from the hills to see the great fair.

To offset this Steel Elephant exhibition, what is the Farmers' Friend Tractor Company doing? In deep shame and humiliation, I am forced to reply that we are doing practically nothing. Our exhibition tractor with the tent and literature and all the supplies has not yet come. And the freight agent has just received word by telegram from Schenectady to the effect that the car containing our stuff has been derailed in a bad freight wreck near that point and that it will not be possible even to start it out of Schenectady until next week.

Consequently our entire exhibit at the fairgrounds consists of a large sign which I have had painted stating that the world's record-breaking endurance test is now going on at Mr. Eben Lockwood's farm just outside the town. I have similar placards all over Marblebury, but what good does it do? Mr. Lockwood's farm is not even on the main road. It is on a side road, about a mile from the fairgrounds, and practically, nobody seems to take the trouble to go out there. If I only had the brass band and the bathing girls we could attract so many people that we would make the Steel Elephant efforts look completely sick. But when the company absolutely turns down all my recommendations, there is not much that I can do. In your telegram you say that our exhibit must be interesting and attractive, but conservative and dignified. I will do my best to carry out the wishes of my superiors. But how can you make anything interesting and attractive if you have to be dignified and conservative?

Yours in deep discouragement,
ALEXANDER BOTTS,
Salesman.

FARMERS' FRIEND TRACTOR COMPANY
SALESMAN'S DAILY REPORT

DATE: SEPTEMBER 15, 1922.
WRITTEN FROM: MARBLEBURY, VERMONT.
WRITTEN BY: ALEXANDER BOTTS, SALESMAN.

Today I had a splendid idea. If I could not get the people out to the endurance test, I could bring the endurance test in to the people. There is a nice grass field just across the road from the main gate to the fairgrounds, and it occurred to me that I could have Mr. Lockwood come in and do the last days plowing on this field. Mr. Lockwood, when I approached him on the subject, was surprisingly agreeable, and agreed to do this if

I could make arrangements with the owner of the field. Unfortunately, however, it turned out that this field is owned by Mr. George Crossman, the president of the Marblebury Fair Association and local dealer for the Steel Elephant Tractor Company. When I approached Mr. Crossman on the subject he was very insulting indeed, and informed me that he would permit no Earthworm tractor exhibit on his land. In addition he gave me other unpleasant information.

"I hear," he said, "that you have been telling it around town that when you finish this foolish endurance test, as you call it, you are going to have a parade and bring your tractor in the main gate of the fairgrounds and drive it around the race track, and I don't know what all besides. Is that right?"

"Well," I said, "I will admit I had thought of doing something of the kind."

"You might just as well forget it," he said. "I am the president of the Fair Association and I won't permit it. Our rules are that all machinery must be brought in the rear gate of the fairgrounds, taken directly to the place of exhibition and kept there until the fair is over."

"But if I take it in the back gate," I said, "and run it right in to that miserable little lot away over in the fence corner, hardly anybody at all will see it."

"Possibly so," said Mr. Crossman unpleasantly, "but that is your hard luck, not mine. And if you get to feeling too downhearted you might cheer yourself up by coming around and seeing the great bathing beauty contest on Saturday afternoon."

"Thank you, Mr. Crossman," I said as I moved away. "I do not care to degrade myself by being present at any such exhibition of sensuous carnality."

This parting bit of repartee cheered me up slightly—but not much. Things are indeed in a bad way. Here I am trying to carry out the company's orders and put on an interesting and attractive exhibition, but I get no cooperation from anybody. The fair authorities won't let me have a parade. They won't even let me come in the front gate or use the ground across the road. The railroad company has done me dirt. Instead of bringing our tractor and supplies to the fair at Marblebury, Vermont, they throw it into the ditch at Schenectady, New York. And my own company won't stand behind me. All they do is tell me that I have to be dignified and conservative. P.T. Barnum himself could not have put on a good show if he had been hampered like this.

The endurance run is going good. Mr. Lockwood is certainly doing his part fine. But there is nobody there to see it. If only the bathing girls hadn't got away from me it would have been entirely different. But as it is, I will probably have to report in tomorrow's letter that the whole business has been a dismal flop at far as publicity and sales are concerned.

Yours,
ALEXANDER BOTTS,
Salesman.

FARMERS' FRIEND TRACTOR COMPANY
SALESMAN'S DAILY RETORT

DATE: SEPTEMBER 16, 1922.
WRITTEN FROM: MARBLEBURY, VERMONT.
WRITTEN BY: ALEXANDER BOTTS, SALESMAN.

In spite of superhuman difficulties and ending discouragement, I am pleased to report that I have not lost my fighting spirit. And today I had another of my brilliant ideas. I was down at the main gate of the fairgrounds, watching the early arrivals driving in with their automobiles. There is so much automobile traffic during the fair that the town police force is unable to handle the traffic all by himself. Consequently various men appointed by the Fair Association act as special temporary traffic officers. These men wear no uniforms, but their official position is indicated by the fact that each one of them wears a celluloid badge bearing the words Marblebury County Fair and a ribbon attached thereto with the words Traffic Officer.

As I stood at the main gate I noticed that one of these beribboned special officers was directing the traffic which was coming in from the state road, and I was greatly interested to observe the docile way in which the average citizen obeys anyone who appears to have authority. All that was required was a simple wave of the hand and the automobile drivers obeyed as automatically and promptly as if the command had come from heaven itself. As I watched this ready obedience it suddenly occurred to me that if I had a badge and ribbon I could direct traffic with the same facility.

With me, to think is to act. I at once called Mr. Samuel Simpson.

"Sam," I said, "do you think you could get hold of one of those badges for me?"

"Sure," said Sam. "I saw a lot of these on the table in the main office of the Fair Association. I could sneak in there and swipe one without any trouble at all."

"Fine!" I said. "You certainly are a good man to have around. Get me one as soon as you can and meet me out on the main road right where the little side road turns off to Mr. Lockwood's place."

As Sam started for the main office, I went to a hardware store in town and had them paint me a sign on a large piece of cardboard. I then hurried out to the crossroads and set up my sign. Road Closed. Detour to the Fairgrounds. A moment or two later good old Sam arrived with the badge and ribbon, which I immediately pinned on the lapel of my coat.

"Sam," I said, "you will now go over to Mr. Lockwood's farm and stop all cars that come along. Tell them that if they wait a little while they will see the thrilling finish of the great world's record tractor run."

"Right," said Sam, and hurried up the little side road.

By this time it was exactly two o'clock in the afternoon, and for the next two hours I made every car that came along detour so as to go past old Mr. Lockwood's farm. As this was the last and biggest day of the fair, and as this was the most important road leading into town, there were a tremendous number of cars. A few drivers tried to argue with me, but as I was completely hard-boiled and was backed up by my badge and by the big detour sign, they soon gave in and did as I told them.

At this point I may as well explain that my sign was perfectly true. The little side road, after passing Mr. Lockwood's place, curved around and came back to the main road very near the gate of the fairgrounds. It really was a detour leading to the fair. And it was also true that the main road was closed—I had closed it myself.

At four o'clock an old flivver automobile appeared from the direction of Mr. Lockwood's farm. When it reached the cross roads it stopped and Mr. Lockwood himself got out.

"What do you think of me as a traffic cop?" I asked.

"You sure are a wonder," said Mr. Lockwood, "and that man Sam is even better. We've got cars parked around our place as thick as flies on a dead horse. We're getting publicity at last."

"Fine!" I said "You may be able to get into vaudeville or Congress yet. How is the tractor running?"

"My son is driving," said Mr. Lockwood, "and he's sailing along just as stylish and handsome as Dewey at Manila. We only have an hour more to go. Hadn't you better be coming over?"

"It is a good idea," I said. "But couldn't we drive around by way of the fairgrounds and check up on what old George Crossman is doing?"

"It's all right with me," said Mr. Lockwood. "Let's go."

I sent three more cars down the detour and then drove off with Mr. Lockwood, leaving the detour sign in place to the hope that it would continue to influence the traffic.

When Mr. Lockwood and I arrived at the fair we found that there was a dense crowd about the Steel Elephant Tractor Company's tent. After elbowing our way for some distance into this mass of humanity we got near enough to see that the bathing beauty contest was in full swing. To my eyes it was a most melancholy sight. The girls themselves were really very easy to look at—especially the little queen of the soda fountain from Hopkins' Drug Store—and I will have to admit that the bathing suits were unusually modest, as bathing suits go. But I was nevertheless greatly distressed to see these beautiful and innocent young women exhibiting themselves for the sake of George Crossman's paltry fifty dollars.

The demoralizing influence of the whole performance was apparent to the behavior of the crowd. Instead of being uplifted and ennobled, as they would have been had I put on the show, they were pressing forward with so much vulgar eagerness that it took Mr. Lockwood and myself at least five minutes to fight and shove our way to the front row so that we could get a good look.

We viewed the nauseating spectacle for five minutes, and then we fought our way back through the crowd, climbed into the flivver and drove out to Mr. Lockwood's farm. Here we found an even greater crowd than at the fair. As Mr. Lockwood had said, I had been a great success as a traffic cop, and Sam had been even better. Furthermore, we had been aided by the fact that Mr. Crossman had advertised his show so poorly that very few out-of-town people had even heard of it. The pasture beside Mr. Lockwood's barn was completely filled with hundreds of parked cars, and the occupants were watching Mr. Lockwood's tractor as it moved slowly up and down the adjoining field nearing the end of its great record.

At once I climbed upon the roof of the cow shed and delivered one of the most eloquent and successful speeches of my entire career. In ringing

At once I climbed upon the roof of the cow shed and delivered one of the most eloquent and successful speeches of my entire career.

tones, I told these people that they were about to witness the end of a stupendous epoch-making performance—seven days and seven nights of the most grueling labor which any machine made by man had ever passed through. I pointed out that throughout all this terrific test the Earthworm had never faltered. If necessary, it could go on indefinitely.

"Obviously," I said, "the Earthworm is the only tractor fit for the stalwart farmers of Vermont. Every one of you should own one of these wonder machines. They are sold on easy terms. And as soon as this record-breaking run is finished, Mr. Samuel Simpson will circulate among you and write up your orders."

Note: I had decided to let Sam do this clerical work in order that my time might be free for the more inspirational work of speech making.

At five minutes before five I finished my address amid thunderous applause, and at exactly five o'clock by the farm agent's watch Mr. Lockwood's son drove his tractor up to the barn and shut off the motor. I was the first one to grasp him by the hand. Then the large crowd—which I had so successfully diverted from the main highway and from Mr. Crossman's shameless saturnalia—gathered around him, slapped him on the back and cheered and congratulated him in a manner that was splendid to see.

I was somewhat surprised to find that the crowd also insisted on congratulating me in the most friendly way. For some reason they seemed particularly interested in the fact that I was wearing an official badge of the fair. At least a dozen different men came up to me, looked at my badge and said that it was most appropriate, and that it fitted in with my speech making very well indeed.

I thanked them all most politely and circulated about, talking and chatting with first one group and then another. And I am pleased to report that wherever I went I was greeted with good-natured smiles from one and all. After half an hour or so the crowd melted away and I was able to check up with Samuel Simpson on what he had been doing.

Sam had done well. He had moved through the crowd accompanied by Mr. Lockwood, who had pointed out to him the substantial farmers who would be most apt to buy tractors. He had talked with these people in his quiet way, and with Mr. Lockwood backing him up in everything he said regarding the success of the Earthworm tractor, he had been able to secure orders for no less than two ten-tons and six five-tons. Later in the evening I learned that the Steel Elephant people, in spite of all their fancy exhibiting, had been able to sell only two machines during the whole fair. So you will have to admit that Sam and I have done rather well—especially when you consider that we had no assistance from any brass band or bathing girls.

Yours proudly,
ALEXANDER BOTTS,
Salesman.

P.S.: I have just been looking over that badge and ribbon which I wore this afternoon. Apparently Sam was in such a hurry that he stole it from the stockjudging pavilion instead of the main office. So I have a good joke on the people who congratulated me and told me that it was so appropriate. Probably they never noticed that the blue ribbon fastened to the badge bore the words First Prize Bull.

BIG BUSINESS

ILLUSTRATED BY TONY SARG

FARMERS' FRIEND TRACTOR COMPANY
MAKERS OF EARTHWORM TRACTORS
WESTERN OFFICE, HARVESTER BUILDING
SAN FRANCISCO, CALIFORNIA

APRIL 19, 1924.

MR. ALEXANDER BOTTS,
BILTMORE HOTEL,
LOS ANGELES, CALIFORNIA.

DEAR MR. BOTTS: We have just received a letter from Mr. Spencer K. Yerkes, President of the Bianca Beach Development Corporation, requesting information about Earthworm tractors. Mr. Yerkes states that he is starting the development of a resort property near Los Angeles and will have to do a great deal of grading. We have written him that you will call on him, and we are depending on you to secure his order for as many tractors as his proposed work requires.

Very truly yours,
J.D. WHITCOMB,
Western Sales Manager.

FARMERS' FRIEND TRACTOR COMPANY,
SALESMAN'S DAILY REPORT.

DATE: APRIL 21, 1924.
WRITTEN FROM: BILTMORE HOTEL, LOS ANGELES, CALIFORNIA.
WRITTEN BY: ALEXANDER BOTTS, SALESMAN.

Your letter came yesterday. I called on Mr. Yerkes this afternoon. And I have every reason to suppose that I am about to put across one of the most important deals that I have ever handled since I first became a salesman for the Farmers' Friend Tractor Company. Mr. Yerkes is a big business man in the largest sense of the word, and it is therefore lucky that you entrusted this job to a man like myself, who is able to handle big things. When I tell you exactly what I have done so far you will realize what a tremendous proposition this is, and you will see that I am handling affairs with great skill and gradually working things around to the point of getting a big order.

I did not call on Mr. Yerkes in the morning. These big business men usually spend the time before lunch in reading mail, dictating letters and similar activities, reserving the afternoon for callers. It is therefore a great mistake for a salesman to call on an important prospect in the morning. Furthermore, I find that I am always in better shape to handle the subtle details of a selling talk if I have slept fairly late and not hurried myself at breakfast.

Accordingly I did not make my call until three o'clock in the afternoon. The office of the Bianca Beach Development Corporation is very large and handsome and is every way worthy of the high-grade business which it carries on. The outer office was large and airy. The young lady at the telephone switch took my card in to Mr. Yerkes and returned in a moment or two to usher me through a small gate, across the large outer room and into Mr. Yerkes' private office.

Never before have I seen such a splendid room. There was a thick Oriental rug on the floor. There was a tremendous curved mahogany desk that must have cost at least a thousand dollars. There were mahogany chairs and mahogany filing cases. In one corner was a small office-type electric refrigerator and water cooler of the latest design. And on the walls were several genuine oil paintings of California scenery and Spanish-type stucco houses. Everything was quiet, refined and richly luxurious. And the whole room seemed to be murmuring "I cost money." Naturally I was very favorably impressed.

Mr. Spencer K. Yerkes turned out to be a lean and efficient looking man between thirty and forty years of age. He shook hands with me most cordially, and I was at once struck by his great natural charm and pleasing personality. And before I left, I came to realize that he is also a person of intellect and imagination—a truly big man, capable of handling big things in a big way.

"Sit down, Mr. Botts," he said. "Have a cigar."

"Thank you," I said, taking one.

"I am a business man; my time is valuable. You are also a business man and your time is valuable. Let us get down to brass tacks at once."

"If you will tell me," I said, "what sort of work you are planning, I will be pleased to recommend the machinery necessary for doing it. Until I know exactly what you are going to do, I cannot talk intelligently."

"Quite right," said Mr. Yerkes. "I have a feeling, Mr. Botts, that we are going to get along very well."

"Sit down, Mr. Botts," he said. "Have a cigar."

"I am sure of it," I said.

"What I am planning to do," said Mr. Yerkes, "is to build a town. I have purchased a tract of twenty-five hundred acres. It includes two miles of ocean beach, and it extends back about two miles into the hills. At present the land is entirely unoccupied, but it is within easy driving distance of Los Angeles, and I am going to make it into the city's most beautiful suburb.

"I have bought water rights up in the mountains and I am going to build a five-mile aqueduct which will give Bianca Beach the finest water supply of any town its size on the coast. I am going to lay out streets and boulevards, parks and golf courses. As the ground is hilly, I will have to do a lot of grading. I will have to move a lot of dirt and I will probably need a lot of tractors to do it."

"When it comes to moving dirt," I said, "the Earthworm tractor is the wonder machine of the century. I have pictures and testimonials here in my briefcase, and I am prepared to prove to you that the Earthworm tractor is exactly the machine you want."

"Never mind all that," said Mr. Yerkes, holding up his hand. "I already know the reputation of your company. In fact, I have already figured out that I will probably want twelve of your ten-ton machines."

"I have my blanks right here," I said, "and you might just as well sign an order for them right away. We can ship them out of Oakland tomorrow."

At this Mr. Yerkes smiled. "You are a fast worker and a splendid salesman," he said, "and it is a pleasure to do business with you. But I am not quite ready to sign an order."

"If there is anything more you wish to know about our tractors," I said, "I would be most happy to inform you."

"No," said Mr. Yerkes, "I have no doubts about your tractors. But I am a very conservative and cautious business man, and I have doubts about the buying public."

"What do you mean?"

"I am undertaking a very large enterprise," he said. "I have already spent two hundred thousand dollars for the land and fifty thousand dollars for the water rights, and I am planning to spend an even five million in development. I have every reason to suppose that the project will be a great success. There will be five thousand lots from a quarter to a half an acre in size. If I can sell these lots at an average price of five thousand dollars each, I will take in twenty-five million dollars. If I sell only half the lots, I will still make a handsome profit."

"It certainly looks like a good proposition," I said.

"Yes," said Mr. Yerkes. "But I am conservative. I do not wish to go ahead until I have tested the reaction of the public. I am therefore offering two hundred lots at the absurdly low price of one thousand dollars each, five hundred dollars down and the rest on easy terms if desired. If I can dispose of this offering, I will have proof that the buying public is in the mood to support my scheme. But until I get this proof, I will not move one grain of dirt. I will not spend one cent in development."

"I should think," I said, "that at that price the lots would go like hot cakes."

Mr. Yerkes' reply showed that he is a deep student of psychology. "You might think so," he said. "But the average man is very dense and matter-of-fact. He has no imagination. When you show him a tract of land that is covered with greenwood and cactus, without a street or a house in sight, he refuses to buy a lot at any price. He is a doubting Thomas. But the man of imagination sees more. With his mind's eye he gazes into the future and visualizes the scene as it will be in a year or two. He sees hundreds of beautiful white stucco houses, with their red tile roofs, set in lovely rose gardens. He sees smooth concrete streets and sidewalks, a luxurious country club, a velvety green golf course, and crowds of bathers swimming in the glorious ocean surf or wandering up and down the pure white sands of the beach. The man with imagination buys his lot now for only a thousand dollars. Two years hence the common man will pay five or ten times as much for his. Would you like to see the plans and pictures of the development?"

"Certainly," I answered.

Mr. Yerkes then spent half an hour showing me maps and architect's drawings, and by the time he finished I will have to admit that I was completely sold on his proposition. He is going to have a five-hundred-thousand-dollar country club that will be as magnificent as anything anywhere in the West. And the plans for the new hotel, the Pompeian swimming pool, the Florentine fountain and the civic theater indicate that they will be artistic masterpieces of the first magnitude.

"It almost makes me want to buy a lot myself," I said.

"Evidently," said Mr. Yerkes, "you are one of the people with imagination. But I am not trying to sell you a lot today. However, I would like you to come out and look at the property. I am taking out a party of prospective buyers tomorrow. I would like you to come along, look over the ground, and check up on the grading and dirt moving work which I am planning. With your expert knowledge, you ought to be able to tell me whether I am right in my estimate that twelve ten-ton tractors is what I need."

"Nothing would give me greater pleasure," I said. "I am glad to help you in any way that I can."

"Splendid!" said Mr. Yerkes. "Meet me here tomorrow morning at nine o'clock."

"I will be here," I said. "Good afternoon."

Then—as I am always very careful not to trespass upon the time of a busy executive—I took my departure at once. I have described my call on Mr. Yerkes very fully so that you can see what a very big man he is and what a very big order he is going to give us. With a man like Mr. Yerkes, it does not pay to be in too much of a hurry. But I am going to camp on his trail until these two hundred lots are sold and I get his order for the twelve ten-ton Earthworms.

Yours,
ALEXANDER BOTTS,
Salesman.

Farmers' Friend Tractor Company
Salesman's Daily Report

Date: April 22, 1924.
Written from: Los Angeles, California.
Written by: Alexander Botts, Salesman.

This has been a thrilling day. And when I relate everything that I have done you will see that I have been right up on my toes all the time, that I have handled things in just exactly the right way and that on account of my efforts we are likely to get an even bigger order from Mr. Yerkes than I had hoped. In fact, it would not surprise me at all if we sold him twenty ten-tons and about six five-tons.

At exactly two minutes before nine this morning. I walked into the office of the Bianca Beach Development Corporation. I sent in my card, and at exactly nine o'clock Mr. Spencer K. Yerkes himself came out and greeted me.

"Good morning, Mr. Botts," he said, "I see you are punctual. You are a man after my own heart. We will start at once."

"I am ready any time you want to go," I replied.

"Good!" said Mr. Yerkes. "I am putting on a real selling drive today. I have advertised in the papers and I am sending out several busloads of prospective buyers. Besides this, I have gathered in three very important prospects, whom I will take out in my private car. These men are all reasonably wealthy and I hope to sell them a dozen or more lots apiece. I am expecting that you will ride along with us."

"I should feel highly honored," I replied politely.

"Very good," said Mr. Yerkes. "And there is one thing more. I take it you are pretty well sold on this little development project of mine?"

"I think it is a wonderful thing, Mr. Yerkes," I said. "I am sure it will succeed."

"Splendid!" he said. "Then I shall probably call on you to help me a little in my selling campaign. I can tell that you are a natural salesman, and it wouldn't hurt things at all if you were to talk up this proposition of mine with these prospects. Furthermore, I would be very glad if you would tell them that I have bought a lot of tractors from you and that the work of grading is going to start at once. It is just as much to your interest as mine to sell these lots. The sooner we get them sold, the sooner you will get your order for the tractors."

"That's right," I said. "You can be sure that I will do everything in my power to help you."

"Thank you," said Mr. Yerkes. "I know I can count on you."

He then took me into his private office and introduced me to the three important prospects. They were ordinary looking, uninteresting, middle-aged business men.

"I want you to meet Mr. Alexander Botts," said Mr. Yerkes. "He is the representative of the Farmers' Friend Tractor Company, and he has just sold me twenty ten-ton Earthworm tractors."

Note: As you can imagine, these words fell upon my ears with a most pleasing sound. The day before, Mr. Yerkes had spoken of only twelve but now he had evidently changed his mind and was thinking of twenty.

After we had shaken hands all around, Mr. Yerkes served us some refreshments out of his little trick electric refrigerator, and we then went down and climbed into Mr. Yerkes' car.

It was a splendid eight-cylinder Italian creation, as fine as any I have ever seen. It must have cost at least ten thousand dollars, and it caused my esteem for the owner to mount even higher than ever. Mr. Yerkes is quite evidently a man who considered that the best is none too good for him.

We were soon rolling smoothly along through the Los Angeles traffic. Mr. Yerkes sat at the wheel and beside him was one of the three prospects. The two others sat with me in the rear seat. The man beside me was a rather stolid German looking person by the name of Joseph Schwartzberger. I at once engaged him in conversations and gave him a very good sales talk on Mr. Yerkes' real estate proposition. In doing this I was, of course, indirectly helping myself and the Earthworm Tractor Company; for, as Mr. Yerkes had pointed out, the sooner these lots were sold, the sooner we would get the order for the twenty tractors.

Unfortunately, Mr. Schwartzberger seemed very dumb and unreasonable. He kept gazing at the scenery as we drove along and seemed to take very little interest in my glowing descriptions of the splendid development work which Mr. Yerkes was going to do. His mind kept wandering off the subject, and he kept talking about orange growing—which had nothing to do with the Bianca Beach development at all, as the ground there is too rough for orchards. If I had let him ramble on, probably we would have talked about nothing but orange growing. However, each time he brought up the subject I very skillfully changed the discussion to the consideration of the new Bianca Beach five-hundred-thousand-dollar country club, the Pompeian baths or some other feature of Mr. Yerkes' project.

It took us about an hour to reach Bianca Beach, and here I got a very pleasant surprise. Mr. Yerkes had certainly put on his selling drive in splendid fashion. For a quarter of a mile the main road was lined with rows of American flags, flapping bravely in the breeze. There were several dozen enormous billboards announcing the sale of lots in the new Bianca Beach development. And artistically placed amid these billboards was a large tent containing pictures and descriptive literature, and also a large plaster-of-Paris relief map of the project, showing Bianca Beach as it would look when completed. This map was a veritable work of art. The ocean was painted blue, the beach white and the hills green. And there were hundreds of little model houses stuck around amongst thousands of little artificial trees and shrubs. Just outside the tent was a large brass band of at least fifty pieces, which, as soon as we arrived, struck up The Star-Spangled Banner.

There were a dozen or more salesman and other assistants preparing for the crowds which were expected later. Mr. Yerkes spent ten or fifteen minutes checking up in his efficient, businesslike way to make sure that all arrangements had been attended to. Then he took the three important prospects and myself for a short walk to see the principal points of interest in the future town.

The property looked very much as Mr. Yerkes had described it to me in the office. There was a beautiful wide beach washed by the blue waters of the Pacific Ocean. Behind the beach the land was rough and hilly, rising irregularly toward the high mountains five or six miles away. The ground was sandy and covered with bushes and tough grass.

The whole place had been surveyed and the proposed streets had been neatly marked with little white painted wooden stakes. At the street corners were neat little signs with names such as Rose Lane, Desdemona Boulevard, Nightingale Road, Delphinium Drive, and so on. I was interested to learn that Mr. Yerkes had selected these names himself—which shows that he has the soul of a poet as well as the mind of a business man.

As we looked around, Mr. Yerkes showed us the plans and pictures of the future improvements, and described them as vividly that I could almost see them rising up to all their beauty and grandeur from the sandy waste lands. Mr. Yerkes must have been right when he said that I was evidently a man of imagination. The three important prospects, however, seemed to be very unresponsive, and this was particularly the case with Mr. Schwartzberger. It took his dense intellect at least ten minutes to comprehend what Mr. Yerkes meant by the Pompeian swimming pool.

Apparently Mr. Schwartzberger had never heard of Pompeii and missed entirely the significance of the word "Pompeian."

During the course of our walk Mr. Yerkes had me check up on the grading which was to be done in putting through his street program. Owing to the rough hilly nature of the land, there will be a tremendous amount of cut and fill work, and a tremendous amount of dirt will have to be moved to make the park and the eighteen-hole golf course. Besides this, there will be a great deal of dirt moving in connection with building a dam back in the hills, making an open ditch to bring the water down to the reservoir just above town, digging the smaller ditches for the water supply mains and the sewerage system. As Mr. Yerkes was insistent that all this work be completed before the end of the year, I told him privately that I thought he would need about six five-tons as well as the twenty ten-tons he was contemplating purchasing.

"Well," said Mr. Yerkes, "if I need that much machinery I will have to get it. We are doing a big thing here in a big way, and we can't afford to economize on equipment."

Naturally I was overjoyed to hear him say this. And I was also very much pleased a little later when he told the three important prospects very positively and without any reservations that he was buying twenty ten-tons, six five-tons and a lot of other miscellaneous equipment.

One of the prospects—a man by the name of Smith—seemed particularly impressed with this fact.

"That piece of information," he said, "is the one thing necessary to make me decide to buy. A man is naturally leery about buying lots before the property has been improved. But now that I know you are actually getting this large amount of machinery, I feel very much reassured, and I have every confidence that the improvements will go through according to schedule."

Mr. Schwartzberger, however, was skeptical. On the way back to the tent he spoke to me privately.

"Are you really selling this man that much machinery?" he asked.

"Certainly," I replied. "That is, I am selling him the twenty-six tractors and he is getting the other equipment from the respective manufacturers."

"Have you got your money yet?" asked Mr. Schwartzberger.

"Really, sir," I replied, "I do not feel that I ought to discuss all the details of my customer's financial arrangements, but I might say that naturally the Farmers' Friend Tractor Company does not expect payment until the machines are delivered—and that will not be until next week."

"Well," said Mr. Schwartzberger, "I hope you get your money. You know practically all these real estate fellers are crooks. I very much doubt if I'll buy any lots here. The proposition sounds good, but there must be a catch in it somewhere. And by the way, sometime I want you to tell me whether your tractors are any good for work in orange groves."

"Why talk about oranges groves now?" I said. "The best thing for us to do here is to find out all we can about this wonderful development scheme. I only wish I could get you to take a little more interest. I really believe, Mr. Schwartzberger, that if you do not buy as many lots as you are able to, you will be throwing away the opportunity of a lifetime."

"Maybe so," he said, "but I hate to take a chance."

By this time we had got back to the tent, which was now full of people. Four large busloads of prospective buyers had arrived from Los Angeles and a great many more had come in their own cars. Mr. Yerkes at once took charge of these people and led them around the same route which we had taken. At each important point he stopped and gave a short explanatory talk. When this tour of inspection was finished, a light lunch was served to everyone to the tent, while the band played stirring music.

After lunch, Mr. Yerkes mounted a small platform and delivered one of the most remarkable orations I have ever heard. I am a pretty good talker and a pretty good salesman myself, but from now on I am going to take Mr. Yerkes as my model and master. The man is a wonder. He is a super salesman and a modern Demosthenes.

He started his talk on a very high plane. He described in great detail and with a wonderful warm flow of language the magnificent building operations which would soon commence. He painted such a convincing and vivid word picture of the lovely Pompeian swimming pool that he made us all feel as if we were standing beside it ready to plunge into its cool and inviting waters.

He then shifted to the golf links and the five-hundred-thousand-dollar country club, and at once we were transferred to the velvety greens with

"Have you got your money yet?" asked Mr. Schwartzberger.

our golf clubs. And a moment later we were sitting on the spacious veranda sipping cool drinks and gazing out over the lovely green landscape. By his consummate artistry and mastery of words he took us for a walk down the beautiful palm-lined avenue and past the cozy white houses set in gardens of fragrant roses. He showed us the cool and delightful park, the crowds of bathers on the beach and the happy throngs going to the moving picture show in the civic theater.

And then he suddenly became quite businesslike. With hard facts and remorseless logic, he drove home the great truth that the people who bought lots today would be able to sell them in a year or two for ten times what they had paid for them.

"But," he said, "I hope that you will not want to sell. I hope that you will all remain in this paradise on earth. Bianca Beach is to be a city of homes, where you may all dwell in peace and contentment with your gracious wives and your darling children. Picture to yourselves what this place will be like in only one or two short years. Crowds of happy youths and maidens will be sporting in the surf or cleaving the waters of the Pompeian swimming pool. Young and old will be spending many happy hours on the golf links in the glorious California sunshine. Others will idle away the balmy afternoons under the shade of the palms in the park. There will be splendid schools, churches and a library, moving pictures, high-grade stores of all kinds, beauty parlors, barber shops, garages and filing stations. But greatest of all will be the homes, where happy families will dwell amid all the most modern conveniences, making use to the fullest extent of our up-to-date electric light service, our copious water supply and our splendid sanitary sewerage system. It is a beautiful thought, men and women—a beautiful thought.

"And in conclusion I wish to state that the young man at the table by the entrance to this tent is prepared to sell you as many lots as you desire to take. And I would advise you not to delay. This is a limited offer at an absurdly low figure, and if you do not act at once your opportunity to profit by these prices will be gone forever. I thank you."

I have repeated Mr. Yerkes' exact words, as near as I can remember them, so that you can see that I am profiting by this opportunity to improve myself. The basic principles of all selling are the same, whether it be real estate or tractors, insurance or anything else. I am making a thorough study of Mr. Yerkes methods. I have always been pretty good myself, but in Mr. Yerkes I recognize a real genius in the higher art of selling. And by taking him as my ideal I confidently expect to become an even greater tractor salesman than I have been in the past.

As soon as Mr. Yerkes had finished his masterly address the people—several hundred of them—began crowding around the table where the lots were being sold. The effect of Mr. Yerkes' talk was so stupendous that it took fifteen minutes to get the crowd under control and make them line up in an orderly manner, and it was two hours before everyone could be attended to. There were a few people, of course, who bought nothing, but most of them bought at least one lot, and many bought more. I, myself, have no use for a lot, but I was so impressed by the investment possibilities of the scheme that I decided that I must have one. Upon consulting my check book I found that I had only three hundred and twenty-five dollars available, but Mr. Yerkes, as a special personal favor, consented to let me take a lot with only three hundred dollars instead of five hundred dollars as a down payment.

Of the three important prospects who had come out in Mr. Yerkes' private car, two of them purchased thirty lots apiece, but Mr. Schwartzberger remained stubborn and bullheaded to the end. In view of the fact that Mr. Schwartzberger had been riding in Mr. Yerkes' car and had eaten his food, it seemed to me that this was a very small way for him to act.

Some people are so mean and so suspicious, and have so little trust in human nature, that they won't do any business at all for fear somebody will slip something over on them. If you took a guy like this Schwartzberger to the bank and got out a five-dollar gold piece which was guaranteed by the cashier to be genuine and offered it to him for fifty cents, he would probably turn it down unless you agreed to give him his money back if he wasn't satisfied; and at that, he would probably want ten percent off for cash.

After the last customer had been satisfied, Mr. Yerkes gathered up the records and receipts, left his assistants in charge of the tent and equipment, started up his elegant motor car and drove us all back to Los Angeles. On the way he made a last attempt to get Mr. Schwartzberger interested. But even a master like Mr. Yerkes is powerless in the face of complete stupidity.

Mr. Schwartzberger had various absurd and evasive replies. First of all he said he would make no down payment until he had the actual deed to the property, with a title abstract and an insurance policy from some reputable title guaranty company. Mr. Yerkes pointed out most reasonably that this procedure would be impossible in this case, as the sales had to be made on the spot and that it would take several weeks of clerical work before the final papers could be made out. In the meantime a signed receipt of the Bianca Beach Development Corporation was sufficient proof that the sale had been made.

Mr. Schwartzberger was still stubborn. He admitted that Mr. Yerkes' point was well taken, but he then proceeded to run in a lot of foolish technical stuff by saying that he would be willing to take a few lots if the money could be put in escrow—whatever that means. But he would not put up any money outright.

Mr. Yerkes' only reply to this proposition was to state that if a man didn't want to risk anything he couldn't expect ever to make any profits. And there the matter was dropped.

When we reached Los Angeles Mr. Yerkes left me at my hotel. In parting, he told me that he had not yet completely checked up all his sales, but that he had sold enough to make him decide to go ahead. He asked me to call on him at this office tomorrow afternoon to close the deal on the tractors.

And so, before tomorrow's sun has set I expect to bring to a successful conclusion the largest and most brilliant sale in all my years of service as a salesman for the Farmers' Friend Tractor Company.

Yours,
ALEXANDER BOTTS.
Salesman.

FARMERS' FRIEND TRACTOR COMPANY
SALESMAN'S DAILY REPORT

DATE: APRIL 23, 1924.
WRITTEN FROM: LOS ANGELES, CALIFORNIA.
WRITTEN BY: ALEXANDER BOTTS, SALESMAN.

Today I have received two sickening jolts, either one which would have been sufficient completely to discourage most ordinary people. But it is pretty hard to keep Alexander Botts down, and I wish to announce that I am still going strong—or as strong as anyone could under the circumstances.

The first jolt came this afternoon when I called at the office of the Bianca Beach Development Corporation. In the outer room I found a small group of nervous and excited people. The place seemed to be in charge of a rather tough looking man whom I had never seen before.

"Where," I asked this man, "can I find Mr. Spencer K. Yerkes?"

"That is what I would like to know myself," he replied, "and so would all these gentlemen who purchased lots from him." With a wave of his hand he indicated the people who were standing around. "I suppose," he added, "you also bought a lot at Bianca Beach?"

"Yes," I admitted, "I bought a lot. And I am here to see Mr. Yerkes on very important business."

"Mr. Yerkes has gone," said the man, "and I doubt if he plans to return."

"But what has happened?"

"Haven't you read the afternoon papers?"

"No, what is it all about and who are you?"

"I am a United States marshal, and I am temporarily in charge of this office."

"Do you mean to say that there was anything crooked about Mr. Yerkes' business?"

"It looks a little bit that way," said the marshal. "We got a tip yesterday from Mr. Schwartzberger, the big orange grower, who thought there was something wrong about this whole Bianca Beach proposition. As soon as we looked into things we found that Mr. Yerkes did not own the land at Bianca Beach at all. All he had was an option which expired last night."

"But he sold several hundred lots." I said.

"Exactly," said the marshal. "And late yesterday afternoon he cashed all the checks he got on down payments and drove out of town in his car. We think he crossed the border at Tijuana."

"It doesn't seem possible," I said.

"Apparently," the marshal went on, "he owes pretty nearly everybody in town. He hasn't paid for his car or for his office furniture. He owes the band and all the people that worked for him yesterday, and the engineers that surveyed his land, and pretty nearly everybody that had anything to do with him. The main thing we are after him for, though, is using the mail to defraud. I hope we catch him. I'd like to meet him. He must be a very slick talker."

"Yes," I said sadly, "he is quite a talker."

At this point the average salesman would have given way to despair. But I am different. I decided there was no use crying over spilled milk. I at once dismissed the nefarious, low-down, slinky Mr. Yerkes from my mind. I resolved to forget my poor unfortunate three hundred dollars. And I as once turned my active mind onto the problem of what to do next.

"By the way" I asked the marshal, "did you say this Mr. Schwartzberger was a friend of yours?"

"I know him fairly well."

"You say he is a big orange grower?"

"He owns about four thousand acres."

"Can you imagine that?" I exclaimed. "The big bum never told me he actually owned any orange groves. What is his address? Where does he live?"

"Out at Pomona."

"Thanks," I said, and started back to the hotel.

I had already mapped out a plan of campaign. My logical mind had at once grasped the situation. Mr. Schwartzberger had been asking about using tractors in orange groves only the day before. And now it appeared that he actually owned tremendous orange properties. Putting two and two together, I decided that he might possibly buy a few Earthworms. If I couldn't sell tractors to Mr. Yerkes, I could to Mr. Schwartzberger. It was too late to go out to Pomona this afternoon, but I resolved to make the trip first thing in the morning.

When I reached the hotel, the clerk handed me a telegram from Mr. J. D. Whitcomb, Western sales manager of the Farmers' Friend Tractor Company. And when I read it I received the second big jolt of this most disagreeable day. Part of the telegram wasn't so bad. It was all right for Mr. Whitcomb to mention that he had received my yesterday's report. It was all right for him to inform me that Mr. Joseph Schwartzberger is a big orange grower and that the Farmers' Friend Tractor Company and a lot of other tractor companies were after him last year, but failed to get any orders. I was amazed, however, at Mr. Whitcomb's closing words.

"You must be sound asleep. If Schwartzberger asked for information about tractors he must be getting interested at last, and if you had had the brains of a half-wit you would have followed him up. See if you can't wake up and get his order before some other company gets ahead of you."

Why Mr. Whitcomb felt it necessary to send such a message I do not know. But I cannot pass over his remarks in silence. Mr. Whitcomb completely ignores the fact that I was not asleep. I was working on a very big proposition, and I was succeeding. I had handled matters so well that I was right on the point of selling twenty-six tractors—and I would have closed the deal, too, except for the fact that Mr. Yerkes was suddenly compelled to leave town.

Furthermore, I object to Mr. Whitcomb's insinuation that my brains are less than those of a half-wit. As I explained earlier in this report, I had shown great intelligence by finding out all about Mr. Schwartzberger before I ever received Mr. Whitcomb's telegram, and I had already decided to go out to see him.

Mr. Whitcomb need have no fear that any other company will get ahead of me. I will call on Mr. Schwartzberger tomorrow morning. And if it is humanly possible I will get his order for as many tractors as he needs, and considering the size of his properties, that ought to be a good many.

To reassure Mr. Whitcomb, and to show him that he is entirely unjustified in his fear that some other company may beat me to it, I wish to state here and now that if Mr. Schwartzberger actually decides to buy any tractors, and if he gives his order to any other company, I will save you the trouble of firing me—I will send you my resignation at once. That is the kind of a guy I am.

Yours,
ALEXANDER BOTTS.
Salesman.

FARMERS' FRIEND TRACTOR COMPANY
SALESMAN'S DAILY REPORT

DATE: APRIL 24, 1924.
WRITTEN FROM: LOS ANGELES, CALIFORNIA.
WRITTEN BY: ALEXANDER BOTTS, SALESMAN.

Today's report will be a very hard one to write. But I will go ahead in my usual straightforward way and I will tell you everything exactly as it occurred.

I reached Mr. Schwartzberger's house in Pomona about the middle of the morning. The old geezer met me at the door, and I will have to admit that he treated me very politely. He said he was sorry I had lost the money I paid out for my lot at Bianca Beach, and it was too bad he hadn't been a little quicker about getting the authorities after Mr. Spender K. Yerkes.

"They'll never catch him now," he said, "He is too smart for them."

Mr. Schwartzberger then took me into his sitting room and introduced me to a young man that was is there.

"Shake hands with Mr. Jensen," he said.

"Pleased to meet you, Mr. Jensen," I said, shaking hands.

"Mr. Jensen is in the same line of business as you," said Mr. Schwartzberger. "He is with the Steel Elephant Tractor Company."

"By the way, Mr. Schwartzberger," I said, "you haven't been thinking of buying any tractors, have you?"

"Yes, I have," he said. "I was even considering getting some of your machines, but when I asked you the other day you didn't seem to know whether they were adapted to orchard work or not. So I decided to do business with Mr. Jensen here."

"You haven't signed an order yet, have you?"

"Mr. Jensen is in the same line of business as you," said Mr. Schwartzberger.

"Yes, I have," he replied. "I am getting fifteen small machines."

"You don't need any more, do you?"

"Not now."

"No chance of you changing your mind?"

"No," he said, "not a chance." And the worst of it is the stubborn old bozo apparently actually meant it. I talked around and argued and pleaded for fifteen or twenty minutes, but there was nothing doing at all. So finally I got up to say goodbye.

Restraining a natural impulse to soak Mr. Schwartzberger in the stomach and put my foot in Mr. Jensen's face, I shook hands most politely with both of them and took my departure.

So that was that. And there is not much more to say at the present time, except that I seem to remember promising you in my yesterday's report that if Mr. Schwartzberger was to buy any tractors from any other company, I would save you the trouble of firing me by resigning.

I see now that I was a bit hasty in making this promise. And I realize—in view of the fact that this Schwartzberger affair is an exception and in no way typical of my habitually successful operations—that my resignation would be a heavy blow to the company. But a promise is a promise. I am not the man to back down. So I hereby tender my resignation as salesman for the Farmers' Friend Tractor Company.

Yours,
Alexander Botts,

TELEGRAM
SAN FRANCISCO CALIF 1105A APR 25 1924
ALEXANDER BOTTS
BILTMORE HOTEL
LOS ANGELES CALIF

YOUR RESIGNATION NOT ACCEPTED MUCH RELIEVED TO HEAR YOU ARE NOT AS GOOD A SALESMAN AS YOU THOUGHT FORGET WHAT YOU SAID IN FORMER REPORT ABOUT TAKING YERKES AS YOUR MODEL STOP IF YOU EVER START SELLING STUFF YOU DON'T OWN TO PEOPLE THAT DONT WANT IT YOU WILL BE TOO GOOD FOR THIS COMPANY AND WILL BE FIRED SURE ENOUGH

JD WHITCOMB
WESTERN SALES MANAGER

SANDY INLET

ILLUSTRATED BY TONY SARG

FARMERS' FRIEND TRACTOR COMPANY
MAKERS OF EARTHWORM TRACTOR
EARTHWORM CITY, ILLINOIS

JUNE 27, 1925.

MR. ALEXANDER BOTTS,
HOTEL MCALPIN,
NEW YORK CITY, NEW YORK.

DEAR MR. BOTTS: On your forthcoming trip into New England we want you to call on Mr. Caleb R. Hubbard, at Hubbardston, Maine. He has just written us that he is thinking of buying a tractor, and we will count on you to get his order for an Earthworm.

If he wants to see a machine in action, you can take him over to Castle Harbor, ten miles from Hubbardston, where our records show that the Maine State Highway Department has a ten-ton Earthworm at work on the roads.

Very sincerely,
GILBERT HENDERSON,
Sales Manager.

FARMERS' FRIEND TRACTOR COMPANY
SALESMAN'S DAILY REPORT

DATE: JULY 1, 1925, 9 P.M.
WRITTEN FROM: HUBBARDSTON HOTEL, HUBBARDSTON, MAINE.
WRITTEN BY: ALEXANDER BOTTS, SALESMAN.

I arrived here early this afternoon. And I am up against a tough proposition. I have had competition before from other makes of tractors and from horses and mules, but this time I have to compete with boats and airplanes.

However, I am going swell. I am getting ready to put on such a wow of a demonstration that it wouldn't make any difference if I was competing against the whole British Navy and a fleet of all metal dirigibles besides. When I explain what I am going to do, you will realize that I am getting better and better all the time.

I hopped off the train at one this afternoon. I checked in at the Hubbardston Hotel, ate lunch, and called on Mr. Hubbard a little before two. Mr. Hubbard turned out to be very intelligent and businesslike, and explained at once what he wanted.

"I own a tract of land on the seashore about ten miles north of here," he said. "At the present time I have a small hotel there called the Seaside Inn. It has been so successful that I am going to build a much larger hotel; which means that I will have to take over a whole lot of building material such as lumber, cement, plumbing supplies, and so forth."

"When it comes to hauling freight," I said, "the Earthworm tractor can't be beat. If it is only ten miles, we could make two or three trips a day."

"The trouble," Mr. Hubbard went on, "is that the place is very inaccessible. If you will step over here, I will show you what I mean."

He led me across the room and pointed to a map which hung on the wall. "Here is Hubbardston—where we are now," he said. "On the seacoast just north of town is Hubbard's Point, which is about five miles wide, and which extends eastward out into the sea about twenty miles."

"Exactly," I said.

"Just north of Hubbard's Point," he continued, "is Sandy Inlet, which is about five miles wide at the mouth, and which extends inland to the west about twenty miles. The Seaside Inn is right here—on a rocky hill just to the north of the mouth of Sandy Inlet. Is that clear?"

"I follow you exactly," I said.

"The inn," Mr. Hubbard went on, "is thus only ten miles north of here as the crow flies, but it is very hard to reach. If you go by sea, you have to sail out around the point. If you go by land, you have to circle around the inlet. Of course, I could go by air, and there is a man in town now trying to sell me a small airplane. I may buy it. I have a field over there big enough to land on, and I could transport the guests of the inn by air very nicely."

"But you couldn't carry much freight on a small plane," I said.

"No," he admitted, "I couldn't."

"How have you managed in the past?" I asked.

"We've been using a rented motorboat," he said. "That means we have to take a fifty-mile trip way out around the end of Hubbard's Point. It takes almost all day. We can't use a large motorboat because we don't have a deepwater landing place at the inn. So, if we have to transport all our building materials on a small motorboat, it will be a very slow and a very expensive process. And whenever there is a storm, we can't make the trip at all."

"I see," I said. "You want to haul the stuff overland. Is there a road?"

"There is a good road," said Mr. Hubbard, "which leads five miles across the base of Hubbard's Point to the south shore of Sandy Inlet. From there you can see Seaside Inn. It's only five miles farther on, but it's on the other side of the inlet, and to get there by land you have to take a fifty-mile drive on very poor wood roads clear around the inlet."

"Why not haul your stuff across the point by tractor," I suggested, "and then take it over the inlet by boat?"

"The inlet is full of rocks," said Mr. Hubbard. "The tide sweeps in and out at about ten miles an hour, and at low tide. It's practically dry—nothing but an expanse of mud and sand, with here and there a bunch of rocks. So it's a bad place for boats. But I thought perhaps we could drive straight across with one of your tractors."

"What!" I said. "With that water running in and out at fearful speed?"

"We would go over when the tide is out," said Mr. Hubbard. "At each low tide we have at least four hours when the sand flats are uncovered. Of course the sand is wet and pretty soft in places. But I have walked out a mile or two at low tide and the sand is solid enough to bear the weight of a man. So if your tractor can run on fairly soft ground, if it can pull a reasonable load in a wagon behind it, if it can make the five miles in less than four hours, and if it is reliable and won't break down and get caught by the tide, I think it will be just the machine I need."

"Mr. Hubbard," I said, "your troubles are over!" And at once I explained to him just exactly how good the Earthworm tractor is, and how it would fulfill all of his requirements. I got out my order blanks and I got out my fountain pen. But, unfortunately, Mr. Hubbard is a very skeptical Yankee. He absolutely demanded a demonstration before he would do business.

"All right, Mr. Hubbard," I said, "if you want a demonstration, you'll have a demonstration."

I hurried back to the hotel. I hired an automobile. I drove across the base of Hubbard's Point to the south shore of Sandy Inlet. Fortunately, it was low tide and I was able to walk out and inspect the sand. It was plenty solid enough for an Earthworm tractor and a wagon.

On the shore was a building with a sign, Down East Canning Company, and out in front on the sand were a lot of men digging clams and taking them into the factory to be canned. These men pointed out the Seaside Inn—a mere speck of a building on the wooded shore far away to the north over the flats.

It looked like a cinch to haul a load over to the inn. An Earthworm tractor, making three miles an hour, could cross in less than two hours. And the tide stayed out for four hours.

Immediately I drove my hired automobile back to town and then ten miles down the coast to Castle Harbor, where I found the state highway department's ten-ton Earthworm tractor pulling a twelve-foot-blade grader along the road. The tractor was in charge of an elderly guy with a walrus mustache, by the name of Andy Meiklejohn. After a long discussion, Andy agreed to drive the tractor to Hubbardston early tomorrow morning and work for me one or more days at a flat price of thirty-five dollars a day. Whether the state highway department will get this thirty-five per day or whether Andy will knock it down for himself, I do not know. And I don't know that I care.

After arranging for the tractor, I went back and called on Mr. Hubbard.

"Mr. Hubbard," I said, "I have just got hold of an Earthworm tractor. I am going to drive it across the sands of Sandy Inlet tomorrow. I want you to have a wagon loaded up with at least five tons of building material for me to drag along. And I hope you can come yourself."

"Fine!" said Mr. Hubbard. "I'll have them load up a wagon this afternoon at the lumberyard. But I can't go with you myself. I have arranged to fly to the inn day after tomorrow morning with the man who is trying to sell me the airplane."

"I will probably see you over there then," I said. "Where can I get an exact timetable of the tides?"

"You had better see Captain Dobbs. He owns the motorboat which I have been using for trips to the inn. He knows more about the tides in Sandy Inlet than anybody else in town."

"Thanks," I said.

I found Captain Dobbs down on the water front, shining up the brass on his motorboat. I explained exactly what I was going to do, and he told me the morning low tide would be from five-thirty to ten-thirty, and the afternoon low tide from about six until ten. As the morning tide is pretty early, I have decided to go in the afternoon.

I thanked Captain Dobbs, and came back to the hotel, arriving just in time for supper. After supper, while sitting on the porch of the hotel, I got to talking with a gentleman from New York who had arrived on the afternoon train. He was a little guy, with a timid and somewhat harassed look on his face. And he said that he and a party of five others were going over to the Seaside Inn tomorrow.

"The inn has a wonderful location," he said, "between the primeval forest and the sea."

"But it's a hell of a place to get to," I said.

"Yes," he agreed. "The women in my party were so seasick in that little motorboat last year, that they almost refused to come this year."

"You don't have to go by motorboat anymore," I said. I then explained how I was going over by tractor, and suggested that he and his friends ride along on the wagon. The gentleman from New York at once went upstairs to consult with the rest of his party, and soon returned, saying that they would accept my invitation with the greatest pleasure. I warned him that the wagon would not be luxurious, but he said anything would be better than bobbing along all day in a sickening little motorboat. So it was agreed that we would all meet tomorrow afternoon a little after four.

Thus you see that I have arranged a splendid demonstration. As usual, I am doing more than anyone could have asked or expected. Not only am I going to show Mr. Hubbard that the Earthworm tractor is the best means of hauling freight over to his inn but I am also going to take a load of his hotel guests and thus prove to him that the Earthworm is the best means of transporting passengers.

By tomorrow night I expect to have Mr. Hubbard's order.

> Very sincerely,
> ALEXANDER BOTTS,
> *Salesman.*

FARMERS' FRIEND TRACTOR COMPANY
SALESMAN'S DAILY REPORT

DATE: JULY 3, 1925.
WRITTEN FROM: HUBBARDSTON, MAINE.
WRITTEN BY: ALEXANDER BOTTS, SALESMAN.

I did not send you any report yesterday because I was too busy. And when I say "busy" that is exactly what I mean. After I tell you what has happened, you will see that I have handled everything in my usual competent and efficient manner. And it was only due to treachery of the basest sort from a quarter whence no one would have suspected it, that the situation today is not so bright as might be desired.

Yesterday morning everything started out very propitiously. The weather was fair, sunny and perfect. At about nine o'clock Andy rolled up to the hotel in his ten-ton Earthworm. As I am very careful and thorough about everything, I had him drive around to Smith's garage, where the two of us spent several hours greasing the machine, changing the crankcase oil, filling up with gasoline and water, and checking over the distributor, the breaker points, the valve timing, and everything else we could think of. The tractor had evidently been given very good care, and by the time we had checked it all over it was as near perfect as a machine could be.

After a late lunch we drove over to the lumberyard and hooked onto Mr. Hubbard's wagon. It was loaded with a lot of heavy planks and timbers, on top of which were tied a number of kegs of nails, a lot of picks and shovels and other tools, and a big road plow. We drove onto the lumberyard scales and found that our load weighed a little more than four tons, including the wagon—not much for a ten-ton tractor, but all right for a trial trip over unknown ground.

When we got to the hotel, the gentleman from New York was waiting for us with the other members of his party—whom I had not seen up to this time. It turned out they were all women. One of them was good-looking. By way of baggage, they had four trunks, eighteen suitcases, a lot of bundles, blankets, sweaters, coats, umbrellas, one dog, one canary bird in a gilded cage, and a large box of fireworks intended, I suppose, for the approaching Fourth of July. Luckily, there were no parrots, cats, goldfish or monkeys.

As the four important females swarmed about the wagon, I began to understand that harassed look on the face of the little gentleman from New York. All four of them began telling him where to put the suitcases and other junk, where to have the hotel porter put the trunks, where they themselves wanted to sit, and saw everything was to be done. As they all had different ideas, and as the gentleman from New York was trying to please everybody, he was soon completely dazed.

Accordingly, I took charge of things myself. I had Andy open up the throttle so that the tractor motor—which had been idling quietly—started up with such a roar that the ladies' conversation was completely drowned out. Then I had Andy shut down the throttle very quick, and before the ladies could start up again I told them with brutal directness that unless they kept quiet they would have to travel by motorboat. They kept quiet.

And in my usual decisive manner I directed the loading of the extra cargo. There was room for most of the smaller bundles in the grouser box of the tractor. The hotel porter got some extra rope and we lashed the trunks and suitcases on top of the lumber behind the plow and the other junk. Then I had four females and the gentleman from New York sit on the suitcases, using the blankets and sweaters as cushions. As soon as they were settled, I tossed them the dog and the bird cage.

This left only the fifth lady—the good-looking one—who all that time had been standing around very modestly without attempting to boss anybody. As a reward for this good conduct I allowed her to ride on the wide, comfortable seat of the tractor with me and Andy.

Note: I have described my handling of that party from New York in order that you may see that I am more than a mere salesman. I am an executive. And it would be a good thing to remember this in case there should ever be a vacancy in some of the higher executive positions in the company.

By the time we were ready to start, the four ladies in the wagon had begun to chatter once more. They wanted to know how long the trip would take, and what they could do if it rained, and was the wagon perfectly safe, and so on.

"All right, Andy," I said, "let's go."

Andy stepped on the gas, the motor let out a splendid roar, and we rolled off up the street. And that was the last I heard from the four important ladies for some time.

The noise of the motor, however, was not so loud as to drown out all conversation on the seat of the tractor. And I learned from the young lady beside me that her name was Miss Mabel Cortlandt. She was the niece of the gentleman from New York. The four imposing females were her aunts.

The tractor ran beautifully and we arrived at the canning factory on the south shore of Sandy Inlet a few minutes after six o'clock—just as I had planned. The tide was out—just as Captain Dobbs had said it would be—and Andy drove straight out onto the vast expanse of slimy sand that stretched away toward the wooded shore line and the Seaside Inn, five miles to the north.

I was delighted to observe that the tractor hardly sank in at all. The wagon wheels cut into the soft sand to some extent, but we had a light load and we moved along as nice as anyone could wish.

We passed very close to a bunch of clam diggers from the factory. They looked at us as if they thought we were crazy, and shouted and waved for us to stop, but we had no time to bother with them. And soon they were left far behind.

As I am always polite and aim to make a good impression on everyone, whether they are prospective purchasers of tractors or not, I started in and explained the advantages of the Earthworm to Miss Mabel Cortlandt. And after I had touched on all the high points of this subject, I began to point out the beauties of the sunset. The orb of day was fast going down on our left, lighting up the water-soaked sands far up the inlet with a beautiful golden glow. It reminded me of a song I had once learned, and at once I began to sing it: "Out on the Deep When the Sun is Low." But just as I started, Andy stopped the tractor.

"What time did you say the tide was due to come in?" he asked.

"Ten o'clock," I said. "Why?"

"Look over there," he said. He pointed toward the east. The flat sands stretched out white and bare.

"You see that black line about a mile or so away?" he asked.

"Yes," I said. "It looks to me like the edge of the water."

"It is the edge of the water," said Andy, "and what's more, it's moving toward us all the time. The tide is coming in."

"That's impossible," I said. "Captain Dobbs told me the tide wouldn't come in until ten o'clock."

"And who is Captain Dobbs?" asked Andy.

"He is the man," I said, "who owns the motorboat that Mr. Hubbard hires to make trips over to the Seaside Inn. He is supposed to know more about the tides in Sandy Inlet than anybody else in Hubbardston."

We sat still for a minute more and watched that black line. It was getting nearer, and it was getting nearer fast. Finally, Miss Mabel Cortlandt spoke up.

"This Captain Dobbs probably knows a lot about the tides," she said, "but that isn't all he knows."

"All right," I said, "what else does he know?"

"He knows that if you get across here with this tractor his job with the motorboat will be gone. So he might have decided that he didn't want you to get across."

"Why, the dirty bum!" I said. "You don't suppose he would really act as low-down as that?"

"I don't know," said Andy, "but if you ask me, I would say we had better turn around and head back toward that canning factory. We're only about a third of the way across, and if we hurry we may be able to get back before we're swamped."

I took another look at the black line of water. It was now less than half a mile away, and it was coming fast. By this time the four ladies on the lumber wagon had noticed it too. They were waving their arms and pointing at it and yelling at me.

"All right," I said to Andy, "let's turn around and go back."

Andy started up, made a wide swing over the sand and headed for the canning factory, which was on the nearest high ground in sight.

All this time the dark line of water was coming on with incredible speed. Before we had gone a hundred feet it had reached us. A half a minute later it was way beyond us, racing up the inlet toward the setting sun. Andy put on all the speed he could and the tractor and the wagon went splashing along in swiftly flowing inch-deep water. In almost no time at all the water was two inches deep.

"Do you think we can make it?" I asked Andy.

"I am afraid not," he said. "It looks bad."

And he was right. It looked very bad indeed. In such situations many people would have given up to panic and despair. But with me it is different. I have always noticed that my mind works at its highest efficiency when I am confronted by a great emergency.

With lightning-like rapidity my brain began to analyze the situation. I knew that the tractor could run through water about two feet deep. But if it got much deeper than that it would reach the magneto, and the whole machine would go dead. We were about a mile and a half, or half an hour's drive, from shore. At the rate the water was rising, I knew we could never make it. I remembered that I had noticed the high water marks on the rocks near the canning factory about six feet above the level of the sand. If the water eventually got six feet deep we would be in a bad way. Possibly we could make a raft by tying together some of the lumber on the lumber wagon. But that would have its disadvantages.

And then suddenly I got one of my brilliant ideas. And at once I proceeded to put it into effect. I had Andy stop the tractor and back it up a few inches to loosen the hitch. Then Andy and I got out into the ankle-deep water and unfastened the tractor from the wagon. Next we climbed up onto the load of lumber, moved the trunks, the suitcases, the blankets, the four women, the canary bird, the dog, the gentleman from New York and all the other junk up to the forward end. Then we loosened the chains and ropes that held the lumber and laid a half a dozen long six-by-eight timbers from the rear end of the load of lumber down to the sand. By this time the water was about six inches deep.

Andy then got into the tractor, drove it around to the back of the wagon and started up the timbers. It was a steep climb and the tracks were wet and slippery, but Andy was a splendid driver. And finally—to the accompaniment of encouraging shouts from the young niece, hysterical screams from the four other females, shrill barking from the dog, weak chirping from the canary bird, and silence from the poor little gentleman from New York—Andy got the tractor up on top of the load of lumber.

We lifted the six-by-eight timbers back in place, tightened the chains and ropes so that none of the lumber would be washed away, and made everything shipshape by lashing the trunks, the suitcases and other perishable baggage on top of the big tractor hood as high above the water as possible. Then Andy and I helped the four aunts and the gentleman from New York up onto the tractor. As the seat was already reserved for Mabel and Andy and myself, it was necessary for these other people to perch around on the grouser box and the gasoline tank. One of the ladies held the bird cage in her lap, and another took charge of the pup.

At once I made a short speech.

"Ladies and gentlemen," I said, "I wish to assure you that you are perfectly safe. If you will do as I tell you and keep quiet, no harm can befall you. Our present situation is somewhat inconvenient, I will admit, but it is not due to any negligence on my part. I was treacherously and infamously given false information as to the time of the tides by a man who posed as my friend, but who has turned out to be my enemy.

"When I started to cross these flats, I had every reason to believe that I had ample time to get to the other side. When I discovered that the tide was rising, I attempted to get back to shore. But it was too late. Consequently, I have placed the tractor, as you see, on top of the lumber wagon and I have placed my passengers on top of the tractor. We are perfectly safe. All we have to do is wait until the tide goes out again, when we can proceed on our way."

"But the tide isn't going out," said one of the ladies. "It's still coming in. And it's going to get so deep that it will go right over the top of this machine and we'll all be washed away and drowned. It's terrible! Oh, why did I ever come? I demand that you take us ashore at once."

At once I made a short speech. "Ladies and gentlemen," I said, "I wish to assure you that you are perfectly safe. If you will do as I tell you and keep quiet, no harm can befall you."

"Madam," I said, "I had not finished my talk. If there were any way to take you ashore, I would take you—if only for the sake of getting rid of you. But it can't be done. You will have to stay here, and while you remain you will have to do exactly as I tell you. We are now upon the high seas. Legally speaking, this tractor is now a boat. I am the captain, and under the maritime law of the United States of America I have complete authority over my crew and passengers. If there is any insubordination or disobedience of any kind, I can shoot you or have you tried for mutiny."

As I finished this talk I scowled as darkly as Mussolini himself. And I was gratified to see that the four hysterical females from New York appeared to be completely awed. Andy and the gentleman from New York said I could count on them. And the young niece somewhat surprised me by telling me privately that she was having a swell time, and wasn't it too exciting for words, and she thought my address was wonderful, because it was the first time she had ever seen anyone who could shut up all of her four aunts at the same time.

I thanked her and then borrowed an umbrella from one of the aunts and took a sounding. The water was about a foot and a half deep. Furthermore, the wind was freshening and little waves were beginning to dash against the wheels of the wagon. As the sun sank lower the tide rose higher, and just as the sun disappeared the water reached the bottom planks of our load of lumber. As the darkness deepened, the water crept up farther and farther. The wind blew in stronger and stronger from the sea, and the spray from the breaking waves began to drive over the top of the lumber. The aunts, although I had them too much awed to make a disturbance, nevertheless kept up a continuous chattering. One of them suggested that if we could signal to the shore, somebody might come out in a boat and rescue us.

"It's a splendid idea," said Mabel. "I'll light off some of these fireworks. Maybe they'll send a boat, or maybe they'll shoot us a line and we can all go ashore in a breeches buoy."

And right away she climbed out over the suitcases on top of the hood, pulled out the box of fireworks, and amused herself for an hour or so sending up rockets and shooting off Roman candles. But nobody came out from the shore. Probably nobody saw us. Or if they did, they thought we were just having a premature Fourth of July celebration.

At ten o'clock the waves were washing right over the top of the lumber and we all began to get pretty anxious. The current was strong. It was still flowing in from the sea. There was a pale moon, but it was a dark night just the same. It was cold and none of us really knew how high the water would rise before it started down again.

"When I studied geography," one of the aunts said, "I was taught that the tide in the Bay of Fundy rises seventy feet. What if it gets that high here?"

"It won't get that high," I said.

"What if it rises only half that far?"

"It won't," I said, although I wasn't sure. "And what is more, I don't want any more pessimistic remarks like that out of anybody."

At eleven o'clock the water had risen at least another foot, the wind was still strong, the waves were slashing against the side of the tractor at a great rate and the spray was dashing in onto the floor in front of the seat.

"If it comes much higher," said one of the aunts, "we're lost. And I think it is time you did something, Mr. Captain. This lumber is the only chance we have of saving our lives. But as long as this heavy iron tractor is on the top of it, holding it down, it can't do us any good. What you ought to do is run the tractor off of here while there is still time. Then the lumber will float up to the surface and we can use it as a raft."

"Not on your life," spoke up Andy. "This tractor belongs to the Maine State Highway Department and I am responsible for it."

"And what is a tractor," asked the lady, "as compared to our precious human lives?"

At this point I decided to end the discussion. "The tractor will stay where it is," I announced very decisively, "and this discussion will cease at once. If you people don't shut up, I will have you prosecuted for mutiny, lese majesty, and piracy on the high seas."

They shut up. At half-past eleven it looked as if the water was going down. And at midnight we began to see the uppermost planks of the lumber under the tractor. We knew then that all was well.

And the next two or three hours were really not bad at all. Mabel and I climbed out over the trunks and suitcases and sat on top of the radiator at the extreme front end of the tractor and admired the stars and the moonlight on the waves. We had one interruption when one of the aunts protested that I was getting too familiar with her niece—which was absurd, because I was only protecting her from the cold and the damp sea air. After I had threatened to put the aunt in irons for the rest of the voyage, she quieted down.

Gradually the water sank lower and lower until finally, just as the sky to the northeast began to brighten with the dawn, I looked down and saw wet, shiny sand all around us.

Andy and I put the big timbers in place at the back of the wagon. Andy backed the tractor down onto the sand and drove around and hitched onto the wagon once more. As far as I was concerned, I was ready to go on to the Seaside Inn. And Andy and the gentleman from New York and his niece were game. But the four aunts set up such a roar and demanded so loudly to be taken back to the nearest dry land that I decided the easiest thing to do would be to humor them.

Consequently, we set off full speed for the canning factory, and in about half an hour we had almost reached the shoreline. I had decided to dump my passengers at the factory, where they could telephone for a taxi to take them to town, and I was then going to turn right around and head for the Seaside Inn, which I was certain I could reach before the tide came in again.

But about a hundred yards from shore we ran into a patch of mud which was much softer than the sand. The tractor stayed on top very well, but the wagon began to sink in so deep that I was afraid it would get completely stuck.

"Whoa!" I said to Andy. "I think we had better unhook the tractor, drive it around and hook onto the rear, so we can pull the wagon backward out of that sand. As soon as we get the wagon onto the firm sand we can hook on in front again and circle around this soft spot."

"All right," said Andy, "it's a good idea."

Unfortunately, it was not a good idea. We had no trouble hooking on to the rear of the wagon, but as we pulled it backward we must have backed the nut off the end of one of the axles. When we were just about halfway out of the mudhole the left hind wheel came off, the left hind corner of the wagon dropped down, and the four ladies, the gentleman from New York, the four trunks, the eighteen suitcases, the road plow, the nail kegs, the dog and the canary bird all slid off gently but firmly into the mud. It certainly was lucky that Andy, Mabel, and I happened to be on the seat of the tractor.

For some reason or other, the four aunts seemed to blame me for this accident, although it was nothing that I could have foreseen or prevented and was obviously due to faulty design in the wagon. They shook their umbrellas at me and told me exactly what they thought of me—which apparently was not much. After what they said, it would have served them right if I had let them waddle ashore through all the mud. But I am naturally chivalrous and kindhearted, so I had Andy make several trips with the tractor and carry them and their belongings over to the canning factory.

You might have supposed that this kind treatment on my part would have earned their gratitude. But such was not the case. They all trooped into the factory—which was not locked, although the workmen had not yet appeared—and one of them called up Mr. Hubbard on the telephone. She told Mr. Hubbard to come out and get them at once, and she said that they had been thrown in the mud, insulted, kidnapped, and half drowned by a crazy tractor salesman. After the telephoning was over, they all stood around and glared at me—that is, all but the gentleman from New York, who was too timid, and his niece, who was too sensible. As there didn't seem to be much I could do for these people, and as some of them did not seem to be enjoying my company, I withdrew and went out with Andy to work over the wagon. After hunting around a while we were fortunate enough to find the nut which had come off the axle.

"If we were on a hard road," said Andy, "and if we had a good jack we could lift up this axle and put the wheel back vary easy. But as it is, I'm afraid we'll have to take off the whole load of lumber."

"I'm afraid you're right," I said.

Pretty soon we saw Mr. Hubbard driving up to the canning factory. He had come in a hurry. At once the four excited females gathered around him, talking fast and furious, and apparently giving him their version of what had happened. They must have poured him out a good earful, because very shortly we saw him coming across the sand like a cavalry charge. Andy and I walked forward to meet him, and he was positively foaming at the mouth.

"This is the damnedest proceeding I ever heard of!" he said. "What do you mean by pulling off such a stunt? You told me you were going to haul a load of lumber over to the Seaside Inn. Instead of which you kidnap a lot of my guests. You take them out into the middle of the bay. You pretty near drown them. You scare them half to death. Then you wreck my wagon and dump them all into the mud. It's an outrage!"

"But, Mr. Hubbard," I said, "you don't understand. I can explain everything."

"I don't want to hear another word," interrupted Mr. Hubbard, "and I don't want any explanations. I don't want anything more to do with you.

I wouldn't take your tractor as a gift. The best thing you can do is get out of town as fast as you can. If you ever even speak to me again, I'll knock your block off."

And before I could answer he turned around and went back to his automobile. The party from New York all piled in. I heard Mr. Hubbard tell them he would send back a truck for their trunks and suitcases. Then they drove off toward Hubbardston.

Andy and I sat down on the shore to consider the situation. I shall have to admit that I was not completely satisfied with the way things had been going. Of course, I was not to blame for the treachery of Captain Dobbs, nor for the unfortunate loss of the wheel from the lumber wagon. But I realized, nevertheless, that I was to a certain extent in the wrong with Mr. Hubbard. I knew that if I was to sell him a tractor, I would

They shook their umbrellas at me and told me exactly what they thought of me—which apparently was not much.

have to overcome a certain amount of sales resistance. And I decided that the only thing to do was to take that load of lumber across to the Seaside Inn as soon as possible. This would give me a talking point with which I could once more approach my prospect.

"Andy," I said, "if we are able to get this wagon repaired, are you willing to try another trip?"

"Sure," said Andy.

By this time it was almost seven o'clock and the workmen had begun to arrive at the canning factory. The boss of the clam diggers was very much interested in the tractor and asked me what sort of a trip we had had. I had to admit that it was not so good.

"We saw you starting out last night," he said. "We yelled at you to tell you the tide would soon be in. When you paid no attention, we decided

"What do you mean by pulling off such a stunt?"

you probably knew what you were doing. We decided your machine was probably fast enough to get you across ahead of the tide."

"It wasn't," I said. "Would it be possible," I went on, "for me to hire some of your clam diggers to help unload that lumber, put the wheel back, then reload the wagon?"

"I'm afraid not," he said. "We only have about an hour before the tide comes in, and I will have to keep all hands busy to get out enough clams to keep the factory going until this afternoon."

"Maybe," I suggested, "we could speed up the clam digging a little with our tractor."

"You could try," said the boss clam digger.

"Come, Andy," I said. "Let's see what we can do."

We ran the tractor out to the disabled wagon and hitched onto the big road plow. Then we drove back and forth across the mud flats, plowing big, deep furrows, and in about ten minutes we had turned out more clams—according to what the boss clam digger told us—than twenty men could dig in a whole morning.

The boss clam digger was very much pleased and he let us have a dozen men to unload the lumber, put on the wheel and reload the lumber. Meanwhile, three men with baskets picked up the clams.

When the tide began to come in, a little after eight o'clock, we had our lumber loaded and the wagon and tractor parked beside the canning factory all ready to go. And the canning factory had about three times as many clams as they could have dug in the same length of time by hand. It was a very satisfactory arrangement all around.

The boss clam digger told me that the tide would go out again at about four in the afternoon and that the sand flats would be free from water from then until about eight. As I felt that this information was reliable, I decided to start out for the Seaside Inn at four o'clock. In the meantime, I have been sitting around the office of the canning factory writing this report, and eating great quantities of excellent steamed clams which the boss clam digger was kind enough to offer me.

It is now noon and the tide is almost at its highest point. But before long it will be running out, and as soon as it gets off the flats we shall be on our way. The boss clam digger will mail this report when he goes home to Hubbardston this evening. And tomorrow I expect to send you another report stating that I have successfully demonstrated that it is possible to haul freight across Sandy Inlet. I also hope that either tomorrow or sometime within the next few days I may be able to get hold of Mr. Hubbard and talk him around into a reasonable frame of mind.

> Very truly yours,
> ALEXANDER BOTTS,
> *Salesman.*

Farmers' Friend Tractor Company
Salesman's Daily Report

Date: July 4, 1925.
Written from: Hubbardston, Maine.
Written by: Alexander Botts, Salesman.

My report today will be a short one. A whole lot of things have happened, but it will not take long to tell about them.

Yesterday afternoon at about one o'clock, soon after I had finished my yesterday's report, an airplane went by over the canning factory. It was coming from the direction of Hubbardston and it headed out over Sandy Inlet toward the Seaside Inn, so I knew that it must be the airplane salesman taking Mr. Hubbard for a hop. As the machine went over, I noticed that the motor was missing and spluttering a good deal. But it flew right on until it got more than halfway across the inlet. Then it seemed to hesitate. And finally it glided down gently into the water.

Everybody around the canning factory immediately became very much excited. Because this plane was not a seaplane. It was only a small land machine with wheels on the bottom. The boss clam digger got out a couple of pairs of big field glasses and we stood on the shore and trained the glasses on the plane. It seemed to be about three miles away. Its nose was completely under water, while the tail and the rear edges of the upper wings stuck up into the air. As we looked we saw two men climb up on top of the wings and start waving their arms.

"They're not killed anyway," said the boss clam digger, "and I hope they're not hurt, because we can't rescue them till the tide goes out."

"Haven't you got a boat?" I asked.

"We have an old dory with a motor in it," he said, "but it can't make any speed. And if we went out there now, we'd only get washed out to sea by the tide."

We watched the wrecked plane for about ten minutes. The two men kept up their frantic waving. Then we noticed a motorboat coming in from the sea. It was full of people.

"Good," said the boss clam digger. "That boat has seen their signals. And it seems to be fast enough to buck the tide."

We watched the boat. The tide was evidently pretty strong, but the boat came along steadily. It had almost reached the plane when it suddenly stopped. We stared at it through our glasses until our eyes were tired, but we couldn't see that it moved an inch. It seemed to be stuck.

Two o'clock came. Then three o'clock. All this time, of course, the tide was running out. And a little before four o'clock the sand flats began to emerge.

"All right," I said, "it's time for us to be moving."

Andy and I got into the tractor, and with the load of lumber rolling along behind, we started out across the inlet. Everything went fine. And about an hour later we had reached the stranded plane. The propeller had been broken and the wings slightly damaged when they hit the water, but otherwise it seemed to be all right. Mr. Hubbard and the pilot came walking across the sand to meet us. Neither one of them was hurt.

The last time I had seen Mr. Hubbard he had told me that if I ever spoke to him again he would knock my block off. But for some reason or other he had, by this time, apparently changed his mind. When I asked him if he would like a ride for himself and his friend, and a tow for his machine, he replied most politely and with many thanks that he most certainly would. He and the pilot at once climbed upon the lumber wagon.

"The motor went dead on us," explained the pilot. "We smashed things up a little when we came down. But we can fix her up if only we can get her moved out of here before the tide comes back in."

"I'll get her ashore for you all right," I said. "But first I want to see these other people."

I drove over to the motorboat, which was several hundred yards away, high and dry on top of a small rock. It had apparently hit this rock, and there was a good sized hole knocked in the bottom of the boat. As we drove up I heard a female voice. It was Mabel, the young niece of this gentleman from New York.

"Well! Well!" she said. "If it isn't old Captain Botts himself with his seagoing tractor!"

"Right you are," I said. "This seems to be a regular reunion."

And it was. For there in the boat sat the gentleman from New York, the four aunts, the dog, the canary bird, the four trunks, the eighteen suitcases, all the various bundles, blankets, sweaters and umbrellas. And in the stern sat old Captain Dobbs.

I at once jumped down from the tractor and advanced upon Captain Dobbs, scowling in a very threatening manner.

"Captain Dobbs," I said, in a voice that resembled as closely as possible the tone of my old first sergeant in the Army, "you are the guy who gave me a lot of phony dope about the tides in this inlet. And you have got me in very wrong with Mr. Hubbard. If you will own up, I will forgive you and I will salvage your boat. If not, I will leave it here where the next high tide will wash it around over these rocks and probably knock it all to pieces. What do you say?"

At first the poor old captain was very evasive. But I was completely hard-boiled. And finally he admitted that he had lied to me about the tides. He had all kinds of excuses. He was a poor man. He had a wife and children to support. His motorboat was his only source of income. And if Mr. Hubbard got a tractor, he would lose his job of carrying the stuff to the Seaside Inn and he and his family would probably starve to death. I really began to feel rather sorry for the old bird.

Mr. Hubbard listened to all these explanations in silence. Then he merely said, "This is very interesting indeed," and suggested that we move out of there before the tide came back.

Accordingly, we loaded the four aunts, the gentleman from New York, the four trunks, the eighteen suitcases, the dog, the canary bird, the blankets and sweaters, Captain Dobbs, Mr. Hubbard, and the airplane pilot onto the wagon. Mabel took her accustomed place on the seat of the tractor. We dragged the motorboat carefully and gently off the rocks and hitched it on behind the lumber wagon with a piece of heavy rope. We drove over to the plane and hitched it on behind the boat. And then we started across the sands toward the Seaside Inn.

As we moved along we looked like a regular circus parade. And I am pleased to report I had ample power to handle the four tons of lumber, the heavyweight passengers, the motorboat and the airplane. We arrived at the inn all safe and sound, and just in time for a splendid supper.

After I had finished eating I got ready to launch forth on one of my best selling talks. But I didn't need to. Mr. Hubbard stated that he was through with motorboats and airplanes, and he signed up for a ten-ton tractor without any urging at all. His order is enclosed with this report.

Early this morning, when I started back across the sands in the tractor with Andy, everybody was on hand to wish me goodbye and good luck. Mr. Hubbard thanked me, the gentleman from New York and the airplane pilot shook me cordially by the hand, and the four aunts actually thawed out sufficiently to smile pleasantly. The dog barked and the canary bird chirped. Poor old Captain Dobbs waved to me from the shore, where he was hard at work repairing his boat. And Mabel thanked me for having given her the most thrilling, adventurous and enjoyable time she had had for a long while.

As Andy and I drove out over the sand I reflected sadly that one of the most melancholy things about a traveling salesman's life is the fact that he is continually making beautiful friendships which are tragically broken when he has to move to the next town.

I was considerably cheered up, however, when we arrived at the canning factory. The boss clam digger came out, greeted me most affectionately, and at once signed up for a five-ton Earthworm to be used in digging calms. His order is enclosed.

I am leaving for Boston tonight. And in conclusion I wish to state that I think I have done rather well. I have caused two tractors to be bought where most salesmen could not have sold even one. And I have done far more. I have opened up new markets. And it is my fond hope that in the future we may sell many more Earthworm tractors for the four new uses which I have discovered. First, transporting passengers and freight to inaccessible summer hotels; second, rescuing shipwrecked mariners; third, salvaging disabled airplanes; and fourth, adding to the health and nourishment of the nation by digging vast quantities of clams.

Faithfully yours,
ALEXANDER BOTTS,
Salesman.

THE OLD HOME TOWN

ILLUSTRATED BY TONY SARG

FARMERS' FRIEND TRACTOR COMPANY
EARTHWORM CITY, ILLINOIS

JANUARY 5, 1925.

MR. ALEXANDER BOTTS,
LaSALLE HOTEL,
CHICAGO, ILLINOIS.

DEAR MR. BOTTS: We have been informed that the road commissioners of Smedley County, Iowa, are holding a special meeting on January 9th at the county seat, Smedleytown, to consider the purchase of equipment to remove the snow from the county highways, which have been completely blocked by recent heavy snowstorms.

We have shipped to Smedleytown one ten-ton Earthworm tractor equipped with the new Wahoo Improved High-Power Double Rotary Snowplow. We want you to go there at once to demonstrate the machine and, if possible, sell it to the commissioners. We are sending on our service man, Mr. Samuel Simpson, to handle the mechanical end of the demonstration.

Very sincerely,
GILBERT HENDERSON,
Sales Manager.

FARMERS' FRIEND TRACTOR COMPANY
SALESMAN'S DAILY REPORT

DATE: JANUARY 8, 1925.
WRITTEN FROM: SMEDLEYTOWN, IOWA.
WRITTEN BY: ALEXANDER BOTTS, SALESMAN.
SUBJECT: SMEDLEYTOWN SNOWPLOW DEMONSTRATION

As soon as I received your letter, I started for Smedleytown, arriving here early this morning. And I want to say that I would rather put on a demonstration in this town than in any other place in the country. Very possibly you do not know it, but Smedleytown, Iowa, is the birthplace and early boyhood home of Alexander Botts. This is the first time I have been back to the old home town since I left it at the age of eighteen. All my folks have moved away, but I still have a great many old friends and former schoolmates living here. And it will give me the greatest satisfaction to put on a swell demonstration, and let them see how good I am, and sell them some of the finest snow removal machinery in the world.

There will be competition. I hear that the Steel Elephant Tractor salesman is in town. But he has no machine here, so he can't put on any demonstration, and I am not afraid of those Steel Elephant people anyway.

The Earthworm tractor and the plow arrived last night. Sam Simpson got in this morning. He at once unloaded the machinery, and he has been spending the rest of the day checking it over and getting it ready.

We ought to make a big sensation in this town. Neither Sam nor myself ever operated one of these powerful new rotary snowplows. In fact, we never even saw one before. But Sam has been studying the plow that you sent on, and he is sure he can drive it through any drift in the country. Fortunately, there is lots of snow around here. The streets in town have been partly broken out, but most of the country roads are still blocked. So we have a chance to put on a wonderful demonstration tomorrow.

The County Road Commission meets at nine A.M., and I expect to attend the meeting. While I am giving a preliminary address to the commissioners, I have instructed Sam to drive the plow straight down the main street of the town and on out into the country. An hour or so later the commissioners and myself will follow him in automobiles. And as soon as they see what wonderful work the machine does, they will want to buy it right away.

Although the heavy work will not come until tomorrow, I have not been idle today. I have been getting myself in strong with various citizens of the town. This morning I called on Lemuel Sanders, one of the road commissioners, who used to be my Sunday school teacher. I have changed so much that at first he did not recognize me. But when I told him who I was, he gave me a smile that pretty near split his face.

"Alexander," he said, "I am delighted to see you. And I trust that you are still going to church regularly."

"You bet I am," I replied. "I go twice every Sunday, and I am almost always on hand for Thursday evening prayer meeting."

"Splendid!" he said.

Note: I will have to admit that my statement to Mr. Sanders was slightly exaggerated. But he is such a nice old bird that I just did not have the heart to disappoint him in any way. Furthermore, as he is one of the county commissioners, it is just as well for me to humor him as much as possible.

I also called on a couple of my old high school teachers, who were most cordial. My marks in school were never any too good, but they probably realize that this was due to my original and independent way of thinking, rather than to lack of ability.

Old Doctor Merton, formerly our family physician, welcomed me very warmly. He at once demanded whether I had had my appendix removed yet; and if not, why not. I was deeply touched to see the loving way in which he remembered the symptoms of long ago. I had to admit that I still suffered from occasional mild attacks, but I let him know very positively that I did not intend to have an operation until I had to.

Besides these people, I called on a number of other prominent citizens of the town, including Arthur Myers, who used to be in my class in school and is now a reporter on the *Smedleytown Evening Times-Courier*. I gave him a swell interview, which came out in the evening paper. I am enclosing a copy. You will note that I have very slightly exaggerated my position with the company. And I wish to explain that this is not due to vanity on my part. But it seemed to me that I would make a better impression, and that I would, therefore, be much more apt to sell a tractor, if I let them think that I was a person of even more importance than I actually am.

> Very sincerely,
> ALEXANDER BOTTS,
> *Salesman.*

Clipping from the *Smedleytown Evening Times-Courier*:

ALEXANDER BOTTS VISITS SMEDLEYTOWN
FORMER RESIDENT NOW CAPTAIN OF INDUSTRY
WILL SELL TRACTORS TO COUNTY COMMISSIONERS

Alexander Botts, Vice President in Charge of Sales for the Farmers' Friend Tractor Company, manufacturers of Earthworm Tractors, arrived in town this morning and is staying at the Smedleytown Hotel. Older residents will recall that Mr. Botts was born and raised here, and Smedleytown may well be proud of this local boy who has made good in a big way in the world of finance and industry. Mr. Botts, in addition to holding a high executive position with the tractor company, is a director of more than a dozen banks and trust companies. He is in town for the purpose of negotiating a large sale of snow removal machinery in the county road commissioners.

When Mr. Botts was asked to what he attributed his phenomenal success, he replied, "You can tell the young men of this town that the first rule is hard work, the second is hard work, and all the other rules are the same. My own phenomenal rise has been due to the fact that I have never been a clock watcher. I have always tried to give my employers a little more than they expected."

FARMERS' FRIEND TRACTOR COMPANY
SALESMAN'S DAILY REPORT

DATE: JANUARY 9, 1925, 10 A.M.
WRITTEN FROM: SMEDLEYTOWN, IOWA.
WRITTEN BY: ALEXANDER BOTTS, SALESMAN.
SUBJECT: SNOWPLOW DEMONSTRATION.

This is just a short note to let you know everything is going fine. At nine o'clock I left Sam Simpson all ready to start out with the big snowplow. For the past hour I have been giving one of my best sales talks to the county commissioners. They are now listening to the Steel Elephant man, but his arguments are so weak that I am really sorry for the poor fellow.

I will now slip down to the post office and mail this. Then I will come back and take the commissioners to see the work that Sammy has been doing. And this evening I confidently expect to report that I have closed the sale.

Yours,
ALEXANDER BOTTS,
Salesman.

Farmers' Friend Tractor Company
Salesman's Daily Report

Date: January 9, 1925, 4 p.m.
Written from: Smedleytown Hospital, Smedleytown, Iowa.
Written by: Alexander Botts, Salesman.
Subject: Unfortunate Incident Prevents Complete Success of Snowplow Demonstration at Smedleytown.

I regret to inform you that since my last report the situation has changed. And though I never like to give up hope, I will have to admit that our chances of selling a tractor in this town do not seem quite as good as they did early this morning. In order that you may appreciate the difficulties I have run into, and in order that you may understand exactly why I happen to be in the hospital, I will relate as clearly as I can everything that has happened.

I finished my last report at about ten o'clock this morning. After placing it in the envelope and addressing, stamping and sealing it, I started out from the courthouse—where the meeting of the road commissioners was being held—with the intention of going down to the post office, mailing the report, and then coming back to join the commissioners and take them out to admire the work of the snowplow.

That was my intention, but my plans went astray. I got down to the post office all right, and I mailed the report, but I never got back to the county commissioners.

When I left the courthouse—which is in a small park at one end of town—I had no premonition of disaster. As I walked through the park, I noted that it was a beautiful day—clear and cold, with blue sky overhead, and the white snow sparkling all around in the brilliant sunshine. My heart was gay and untroubled.

But as I left the park and turned into the main street of the town, I was startled to observe that the entire business section was in a state of the wildest excitement. Dozens of people were running back and forth across the street, others were gathering in groups as if to discuss something, and from the upper windows of the houses many heads were thrust forth to see what was the matter.

And there was plenty the matter. The building on the corner housed a drug store—or rather, what had once been a drug store. But now the big plate glass window in front was completely smashed in. And inside there seemed to be hundreds of broken bottles and all manner of other material such as is commonly sold in drug stores—magazines, boxes of writing paper, rubber goods, typewriters, picture postal cards, boxes of candy, toilet articles, and so on. Everything was busted up and thrown around in a way that was horrible to see, and mixed in with the debris were great quantities of snow and ice.

The sidewalk, which an hour before had been shoveled clean, was now covered with snow. But the center of the street, which had held at least a foot of densely packed snow, was now almost completely clean. Across the street was a men's furnishing store. Its front was also broken, and inside were great piles of snow, mingled with a mass of neckties, shirts, pajamas, and I don't know what all.

Stepping out into the center of the street, where the walking was good, I continued on my way. I proceeded for two blocks—as far as the post office—and all the windows of all the stores on both sides of the street were smashed in. People were running hither and thither. Others were at work inside the stores, picking up and arranging the scattered stock, and shoveling out into the street great quantities of snow. For once in my life I was completely bewildered. Never had I seen a more curious, a more extraordinary sight.

After mailing my letter I engaged one of the bystanders in conversation.

"What has happened?" I asked. "What is the cause of this catastrophe? Was it an explosion or an earthquake? Or was it a tornado?"

"Tornado nothing!" said the bystander. "Didn't you see it?"

"See what?"

"That confounded tractor snowplow," said the bystander. At these words I will have to admit that a faint feeling of dizziness came over me.

"Tell me about it," I said. "What happened?"

"You certainly missed it if you wasn't here," said my informant. "I was standing in front of Polsky's Hardware Store when I heard a roaring like an airplane motor. And I looked up and I saw this big machine turning the corner onto Main Street."

"Yes," I said. "Go on."

"It was the darnedest looking machine I ever saw. It had a snowplow with two big paddle wheels, one on each aide. As soon as the guy driving it got well around the corner he threw those paddles into gear some way and they began whirling around something terrible, digging up the snow from the street and throwing it off to each side so hard that it sent it right through every window in sight."

"Holy Moses!" I said. "And he drove right along the whole street?"

"He did. I never seen anything like it—big solid chunks of snow and ice crashing and banging through the windows, and the fine dust making such a fog in the air that you could hardly see anything, and people dodging into alleys, and women screaming, and children crying, and that great big motor roaring. And crash after crash, as them plate glass windows was knocked in. I tell you, it's something I'll never forget as long as I live.

"But the man must have been crazy," I said. "Why didn't he stop? Couldn't he see what he was doing?"

"I don't know," said the man. "That fine snow dust made such a fog that maybe he couldn't see anything."

"Why didn't somebody stop him?" I asked.

"I guess everybody was too busy running away from all that flying snow and ice. You see, the snow on this street was packed down pretty hard, and the machine chewed it out in big, solid hunks which went flying through the air like regular young cannon balls. But the sheriff and a lot of deputies are after this man now, and I guess they'll get him before long."

"Where did he go?" I asked.

"He went right down Main Street and on out into the country." The man pointed down the street in the opposite direction from the courthouse. "Who are you?" he asked. "Are you a stranger in town?"

"Yes," I said, "I'm just visiting here for a short time. Yes, probably a very short time."

"You're lucky then," said the man. "All this is going to make it pretty hard for the merchants, and for the whole town. But, of course, it doesn't make any difference to a guy like you."

"Oh, no, of course not," I said. "This won't affect me at all—not at all."

I continued my walk down the street in the direction away from the courthouse. Pretty soon I was stopped by an old gentleman with a beard. He wore large smoked glasses, apparently to protect his eyes from the glare of the sun on the snow.

"Have they got 'em?" asked the old gentlemen. "Have you heard whether they've arrested them yet?"

"I don't know," I replied politely.

"Well," the old gentleman went on, "if they haven't got 'em yet, they will soon. And I hope they send both of them to state's prison for good long terms."

"What do you mean—both of them?" I asked. "I thought there was only one man driving the tractor."

"There was. He is the mechanic, and they're going to stick him in jail just as soon as they catch him. But the main guy they are after is the salesman that was in charge of the machine. It was him that ordered this mechanic to drive it right through the center of town this way."

"You don't know how you interest me," I said. "So they really think the salesman is to blame too?"

"Of course he is. I'd certainly hate to be in his shoes. But if the sheriff finds him first and gets him locked up, he'll be safe enough. It's a good strong jail."

"What do you mean—he'll be safe enough?" I asked.

"I mean, if they once land him inside the jail there won't be much danger of his getting lynched."

"You don't mean they're talking of lynching him?"

"I'll say they are," said the old gentleman. "Most of the men in this town are so mad they'd kill that tractor man as quick as they would a rabbit. But we got a good sheriff; he would die fighting before he'd let any mob take a prisoner away from him."

"I never seen anything like it—big solid chunks of snow and ice crashing and banging through the windows, and the fine dust making such a fog in the air that you could hardly see anything."

"Well," I said, "that's a comforting thought, anyway."

"Yes," said the old gentleman, "it always gives a town a bad name to have a lynching."

"I agree with you absolutely," I said, "and I certainly hope that nobody will do anything to sully the fair name of Smedleytown."

With these words I took leave of the old gentleman. And as I said goodbye to him, I looked with a certain amount of envy upon his copious white beard and smoked glasses. It occurred to me that similar accessories might be very advantageous to a tractor salesman whose popularity was so decidedly on the wane as mine seemed to be. Unfortunately, I could not hope to grow an adequate beard quick enough to help me in the present crisis. Furthermore, I have had no experience in stage makeup; and false whiskers, unless they are very cleverly devised, are worse than useless. I decided, therefore, to give up the idea of hiding behind a bush.

But smoked glasses seemed entirely practicable. And they would not seem unusual, as many people use them on bright, sunny days. After a few minutes' search I located the pitiful ruins of an optician's shop. The proprietor was busy cleaning the snow out of a broken showcase full of glass eyes, but he stopped long enough to sell me the largest pair of smoked glasses that had survived the wreck.

Then, with my hat pulled down over my forehead, with the glasses covering most of my face between the eyebrows and the mouth and with my coat collar turned up so that it concealed my chin, I sallied forth once more. I was rather pleased with my ingenuity in working out this disguise so quickly. I felt quite sure that even the county commissioners, whom I had just talked to for an hour, would not recognize me. For the time being, at least, I was safe.

But I will have to admit that I was not in what might be called a buoyant or jovial state of mind. Here I had come back to my boyhood home, full of enthusiasm, and prepared to make a splendid and magnificent impression upon my former friends and neighbors. It seemed that I was making an impression all right, but not in exactly the way I had hoped.

A short distance down the street I found a group of several men who were discussing the amount of the damage. It seemed to be the general consensus of opinion that the repair bills would amount to something between two and three hundred thousand dollars.

"Of course," said one of the men, "the company that makes the tractor will have to pay all the repair bills."

With my hat pulled down over my forehead, my collar turned up so that it concealed my chin, I sallied forth once more.

These words fell upon my ear in a most unpleasant way. My mind is still in something of a daze, and I have no idea how all this will come out. But I believe I am correct in assuming that the executives of the Farmers' Friend Tractor Company will be distinctly annoyed if they have to pay two or three hundred thousand dollars expenses for a single demonstration; especially in view of the fact that this demonstration may not result in any sale.

I left the group that were discussing this unpleasant subject of damages, and soon met a man who gave me the latest news from the sheriff's posse.

"They found the tractor," he said. "It was stuck in a snowdrift about a mile outside of town. But the driver had skipped. I guess he knew they were after him, and he decided to do a fadeaway while there was still time. But they're sure to catch him later."

"How so?" I asked.

"The country roads are blocked with snow," said the man, "so he'll have to sneak back here. They've got men all over town—at the hotel, and at the railroad station, and everywhere. It won't be long before they have the driver and that salesman both locked up in jail, where they belong."

"Probably you're right," I said.

As I moved on, my thoughts became gloomier than ever. As you know, I am by nature an optimistic soul. I always look on the bright side of things. But the present grim and stunning disaster did not seem to have any bright side. I could not then, and I cannot now, explain why Sammy Simpson wrecked this town the way he did. I have never been the kind of a man who tries to pass the buck; and I will admit that I myself directed Sammy to run the machine through the center of town. But that was because I had never seen one of these high-powered rotary snowplows in action. I thought that the paddle wheels rotated in a reasonable and civilized manner, lifting the snow gently out to the sides of the road. I never supposed that the thing was a combination of a Swiss avalanche and a volcanic eruption on wheels. But Sammy Simpson has always been one of the best men we have. Why didn't he use a little judgment? Why didn't he look around? Why didn't he see what he was doing, and stop before he ruined the whole region? And when he got stuck out in the country, why did he run away and leave the whole mess on my hands?

If we ever find Sammy, we may get the answers to these questions. But in the meantime, everything is a puzzle to me. The only thing I'm sure of is that Smedleytown, Iowa, today looks very much like Soissons, France, in 1918.

I walked the whole length of the business section—more than half a mile—and every single store on both sides of the street was smashed up in a way that was sickening to see. When I had passed the last wrecked store, I turned up a quiet residence street and strolled slowly along, trying to collect my thoughts and decide on some plan of action. But I could not think of anything.

Before long I found myself in front of a large brick building set back among a lot of trees. As I was vaguely wondering what this building could be, an automobile drew up at the curb beside me. A man started to get out. I paid no attention to him and walked on, but he came running after me and laid a heavy hand on my shoulder.

"Those glasses pretty near fooled me," he said, "but I know you just the same, Alexander Botts."

A feeling of black despair settled upon me. Apparently my disguise wasn't so good after all.

And at this point I will have to stop, as the hospital nurse won't let me write any more. She will see that this report is mailed to you tonight. And in a few days, if I am able. I will write you again and let you know more about this most distressing affair.

Yours,
ALEXANDER BOTTS,
Salesman.

Farmers' Friend Tractor Company
Salesman's Daily Report

Date: January 14, 1925.
Written from: Smedleytown Hospital, Smedleytown, Iowa.
Written by: Alexander Botts, Salesman.
Subject: Final Report of Smedleytown
  Snowplow Demonstration.

It gives me great pleasure to report that I am now sitting up and taking nourishment. For the first time in several days I am strong enough to write; so I will give you the additional facts, as far as I know them, regarding this interesting and curious snowplow business.

In my last report I think I told you of my walk through the main street of Smedleytown, and of my bewildered surprise when I found the whole place looking like *The Last Days of Pompeii*. I think I also told you how I walked up a side street, and how a man got out of an automobile and called me by name. At first, I was sure it was the sheriff. But my fears were groundless. For when I turned around, I saw that it was none other than my old friend Doctor Merton, former physician to the Botts family.

"Take off those glasses," he said. "I want to look at you." I took off the glasses "Alexander," he said, "you look sick."

"As a matter of fact," I replied, "I feel sick."

"And I know what's the matter with you."

"So you have just come from downtown, too?"

"No," said the doctor, "I have just been making some calls out in the north end of town."

"So you haven't been down on Main Street for quite a while?" I asked.

"Not since early this morning."

"Then very likely you don't know what makes me feel sick."

"I do," said the doctor. "It's that appendix. When you left here several years ago you had a very well-developed case of chronic, low-grade appendicitis. And you admitted yourself yesterday that it had been bothering you off and on ever since."

"That's right," I said.

"All this time," the doctor went on, "it has been filling your system with poison. You're a sick man."

"Yes," I said, "I feel bad all over."

"You ought to have that appendix out right away," the doctor said.

"This is one of the best hospitals in the state." He pointed to the brick building in front of which we were standing. "I'll get you a room here any time you want. It's a simple operation, and in a couple of weeks you'll be out again and feeling fine. What do you say?"

Rapidly I considered the situation. There were several courses open to me. I could try to sneak out of town, in spite of all the deputy sheriffs that were looking for me. But I have never been a quitter, and I felt that it was my duty to stick around and do everything I could to smooth out this horrible mess.

I could go back downtown. But I had a distinct feeling that for several days at least the county and town authorities would not be in any mood to listen sympathetically to what I had to say—assuming that I could think of anything to say. It appeared certain that for the present I would accomplish nothing except getting myself locked up in jail, or even lynched.

But here was another possibility. I could enter this excellent hospital, where I could rest my weary body, and where my troubled spirit would find refuge from the turmoil of the wicked world.

Somehow it seemed to me that there was something very attractive in the idea of a pleasant bed, a quiet room, and a gentle, sympathetic nurse to look after me. And I was reasonably sure that if I were a sick man in a hospital the deputies and others would not bother me. And they might even feel sorry for me, so that they would be easier to talk to when I got out again.

"All right, doctor," I said, "let's go."

At first he wanted to delay the operation until the next morning, but I told him it would have to be that afternoon or never. And after he had learned that I had eaten no luncheon, and that I had had only a very light breakfast, he consented. He had several other operations to perform that afternoon, and he saved me until the last, so I had time to write up my daily report. It is now four days later. The first couple of days I was a pretty sick pup, but it seems that the operation was very simple and entirely successful, and now I am feeling fairly good again.

This morning they told me I could have visitors. And a few minutes later in walked Sammy Simpson. I was not pleased to see him.

"You don't belong here," I said. "You're supposed to be in jail."

"Why?" he asked pleasantly.

"You know why," I said. "Why did you do it? After you got started, you should have seen that a high-powered rotary plow is no machine to take through the business section of town that way."

"Exactly so," said Sam. "I tested that plow down by the freight station, and as soon as I saw how it threw the snow around I decided not to go up Main Street after all. I drove right out into the country and spent the whole day plowing the main highways. And the machine worked so well and cleaned up so many miles of road that the county commissioners have bought it. They gave me a check in full payment and I have mailed it in to the company."

"Then you didn't smash up the town after all?"

"I certainly did not."

"Sam," I said, "this ether must have affected my brain more than I thought. Because I seem to have a very distinct memory of walking up Main Street and noticing that every single plate glass window was smashed into very small fragments. Everything seemed so clear and distinct that I could have sworn that it was real. And now it turns out that it was nothing but a dream—just a bad dream"

"Dream nothing," said Sam. "That part of it was real enough. After I got back from plowing the roads, I found that a demonstration Steel Elephant tractor and plow had come in on the morning freight a few minutes after I had left. The Steel Elephant operator had been on some sort of a party the night before, and he was still so lubricated that I guess he didn't know just what he was doing. He went down and unloaded the machine; and you saw what he did with it. He certainly did plenty. They caught him that night, trying to hop a freight out of town. He's still in jail. They arrested the Steel Elephant salesman and let him out later. But I hear they're bringing suit against the Steel Elephant Company for two hundred and sixty-seven thousand dollars."

So that is all I have to report at present. The nurse says I have to stop writing and get a little rest. But I will be all right soon. And now that my appendix is gone, I have a feeling that in the future I am going to be a greater and finer salesman than ever before.

Yours,
ALEXANDER BOTTS,
Salesman.

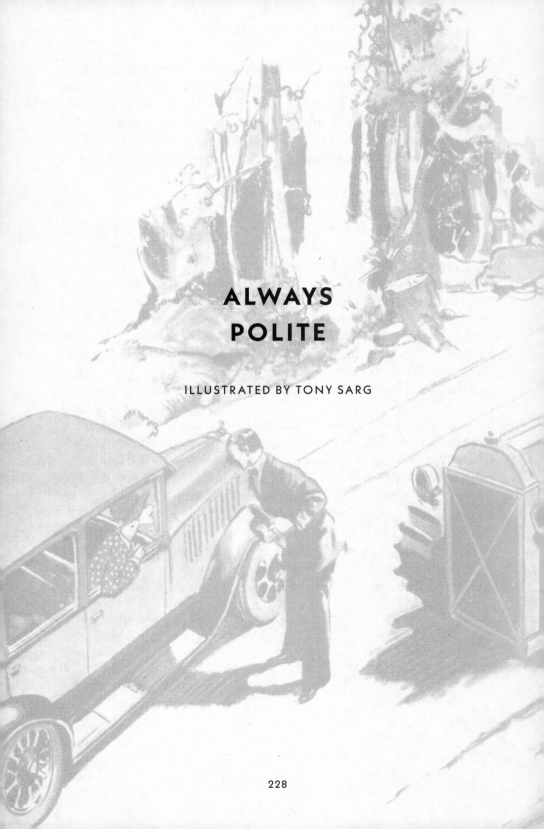

ALWAYS POLITE

ILLUSTRATED BY TONY SARG

FARMERS' FRIEND TRACTOR COMPANY
MAKERS OF EARTHWORM TRACTORS

WESTERN OFFICE,
HARVESTER BUILDING,
SAN FRANCISCO, CALIFORNIA.
APRIL 8, 1926.

MR. ALEXANDER BOTTS,
RITZ HOTEL,
LOS ANGELES, CALIFORNIA.

DEAR MR. BOTTS: As you know, this company has recently decided on a change in its selling organization. The entire country has been divided into districts, in each of which sales will be handled by a local dealer instead of by traveling salesman from the factory. In line with this policy, we have just signed a dealer's contract with the Deane Supply Company, Mr. John Deane, President, of Mercedillo, California.

Mr. Deane writes us that he is going to make a determined effort to sell a tractor to Mr. Harmon Schumaker, also of Mercedillo, one of the largest lumber operators in the state. Mr. Deane is a very good business man, and the Deane Supply Company has for many years done a highly successful business in farm implements. But Mr. Deane has up to this time had no experience in selling Earthworm tractors, and he has asked us to send him one of our factory men to help in putting over the deal with Mr. Schumaker.

We want you to go to Mercedillo at once, cooperate with Mr. Deane in every way possible, and do everything you can to help him put over this sale.

Very sincerely,
J.D. WHITCOMB,
Western Sales Manager.

Farmers' Friend Tractor Company
Salesman's Daily Report

Date: April 12, 1926.
Written from: Mercedillo Hotel, Mercedillo, California.
Written by: Alexander Botts, Salesman.
Subject: Preliminary Operations Preparatory to
    Selling Tractor to Mr. Harmon Schumaker.

I got here this morning and have spent the day gathering information, laying plans for the future, and getting myself in strong with as many as possible of the important people connected with this deal.

First of all I called on our new dealer, Mr. Deane. He may be a good business man, but he is no salesman. He is a gruff old bird, and he is altogether too blunt and direct in his speech. He told me right off that he did not like my looks. He objected to the shine on my shoes, the crease in my pants, and to my natty appearance in general.

"You look like a city slicker," he said. "What it takes to get along with the hard-boiled lumberman here in God's country is a regular he man."

Naturally, I was not favorably impressed by these remarks. I at once sized up Mr. Deane as a guy who would have no more respect for a prospect than he did for me. If a man came in to buy a tractor and Mr. Deane noticed that he was wearing spats, or a wrist watch, or a Phi Beta Kappa key, he would probably throw him out into the street at once. Such tactics are obviously very poor salesmanship.

I saw right away that my job here was going to require a lot of tact. It would be up to me always to remain polite, to pour on the oil, keep things smoothed down, and nurse the negotiations on to a favorable conclusion.

Ignoring Mr. Deane's remarks about my personal appearance, I at once got down to business and asked about the deal in hand. It appears that Mr. Schumaker is at present thirty miles away in the mountains at his sawmill—which is a big enterprise employing five or six hundred men. The logs for the mill are cut away up on one of the higher mountain ranges. These logs are dragged on lines operated by steam donkey engines, from wherever they happen to be, down to a small logging railway. They are then loaded on cars and brought to the mill. If Mr. Schumaker would buy Earthworm tractors to replace these donkeys, he could do the work much quicker and better. Furthermore he could skid the logs much longer distances, and thus save the expense of moving his railroad tracks all the time and building tracks up onto the higher and more difficult slopes. Mr. Deane has one of the new Sixty Horse Power Earthworms on hand, and he has arranged to give Mr. Schumaker a demonstration up in the woods this week.

It might seem that we had here a very simple proposition. But unfortunately, it appears that Mr. Schumaker, the lumber king, is a crabby, disagreeable old gentleman who is very suspicious of new inventions and very hard to do business with.

In fact, Mr. Deane admitted that there was a slight coolness between him and Mr. Schumaker which arose over a business deal some years ago. Since learning this fact, I am more than ever convinced that my main job here is to spread the oil on troubled waters. When the dealer and the prospect are both spoiling for a fight it is especially fortunate that the representative of the Farmers' Friend Tractor Company is a man like myself who is always polite.

We may have some difficulty with a man who is trying to sell Mr. Schumaker a new make of tractor called the Mountain Goat. This machine is especially designed for logging in rough mountain country, but until I see it I cannot form any opinion as to whether or not it will be a serious competitor.

After a short talk with me Mr. Deane went out to attend to some business in town, and I stuck around the office making myself agreeable and helpful to the various assistants and stenographers. Toward the end of the morning a very attractive young lady came in who was introduced to me as Miss Mildred Deane, the daughter of the boss. Following my usual policy of being courteous to one and all, I at once engaged her in conversation. And before long I learned a very important fact. Although the fathers are not any too friendly. It appears that Miss Mildred Deane, the daughter of our dealer, is a former school chum and a very great friend of Miss Sally Schumaker, the daughter of the lumber king. As soon as I heard this bit of news, I invited Miss Deane to have luncheon with me at the best restaurant in the city. I was a little surprised when she promptly accepted, but later I reflected that Mercedillo is a rather out-of-the-way place and probably Miss Deane—charming though she is—does not get very many chances to have luncheon with a young, attractive, and good-looking man of the world.

We had a very pleasant meal, and later went to the movies. In the course of the afternoon I learned that she and Miss Schumaker and some of their friends are starting day after tomorrow on an automobile trip to the Yosemite Valley.

I finally persuaded her to promise that she would drive around by way of Mr. Schumaker's sawmill. She will thus have a chance to see our demonstration of the Earthworm tractor and put in a good word for it with old Mr. Schumaker.

Later in the afternoon I returned to the office of the Deane Supply Company. As Mr. Deane seemed even more nervous and irritable than in the morning, I did not feel it necessary to tell him of the pleasant friendship which had sprung up between his daughter and myself. I stuck strictly to business and made final arrangements for the demonstration.

It appears that Mr. Deane has no mechanic competent to handle the Earthworm tractor, so I am going to drive it myself the thirty miles to the sawmill. As soon as I arrive, I will check over the machine and see that it is in perfect shape to start the demonstration day after tomorrow morning. My next report will come from Mr. Schumaker's sawmill.

Very Sincerely,
ALEXANDER BOTTS,
Salesman.

FARMERS' FRIEND TRACTOR COMPANY
SALESMAN'S DAILY REPORT

DATE: APRIL 13, 1926, 9 P.M.
WRITTEN FROM: SHUMAKER'S SAWMILL.
WRITTEN BY: ALEXANDER BOTTS, SALESMAN.
SUBJECT: TRACTOR DEMONSTRATION FOR MR. SCHUMAKER.

I started out with the tractor early this morning from Mercedillo, and shortly after noon I arrived here at the sawmill. I at once saw Mr. Schumaker, gave him a short but very snappy little sales talk, and arranged to put on a long skidding demonstration tomorrow. Mr. Schumaker is going to be busy all morning, so this demonstration will not be held until afternoon. I have been all over the tractor and it is in perfect condition. With the splendid exhibition of log skidding which I expect to put on, and with the help which Miss Deane has promised to give me, we are practically sure to put over this sale. So far, the man with the Mountain Goat tractor has not yet showed up. He is expected tomorrow, but I am not much worried about him.

> Very sincerely,
> ALEXANDER BOTTS,
> *Salesman.*

TELEGRAM
MERCEDILLO CALIF 9 21P APR 14 1926
FARMERS FRIEND TRACTOR COMPANY
SAN FRANCISCO CALIF

BOTTS AND TRACTOR HAVE BOTH DISAPPEARED STOP FEAR LOST IN MOUNTAINS STOP SEND ANOTHER MAN AND TRACTOR STOP RUSH

> JOHN DEANE

> DEANE SUPPLY COMPANY,
> MERCEDILLO, CALIFORNIA.
> APRIL 14, 1926.

FARMERS' FRIEND TRACTOR COMPANY,
SAN FRANCISCO, CALIFORNIA.

GENTLEMEN: I have just wired you regarding the disappearance of your salesman, and this letter will give you further particulars.

Mr. Botts drove our demonstration Sixty Horse Power Earthworm out to Mr. Schumaker's sawmill yesterday and arranged to give a demonstration this afternoon. Early this morning it appears that he told one of the men at the mill that he was going to take a ride in the tractor to warm it up and see that everything was in good shape. He cranked up the machine and drove it down the small private road which goes from the sawmill through the woods to the main highway in the valley. And neither he nor the tractor have been seen since.

I myself drove out to the sawmill early this afternoon to witness the demonstration. When Mr. Botts did not show up at the appointed time we at once instituted a search. Some of Mr. Schumaker's most experienced woodsmen tried to trace the tractor, but unfortunately there had been a brisk shower after Mr. Botts left, and the tracks left by the machine were almost obliterated. It was impossible to tell whether Mr. Botts had reached the state road or had turned off somewhere into the woods. People living along the state road report that they saw the tractor go by yesterday, when Mr. Botts drove it out to the mill, but they saw nothing of it today. Therefore we fear that Mr. Botts drove into the woods—possibly to try out the tractor over rough ground—and got lost. Searching parties, with lights, are combing the woods tonight, and the search will be continued until Mr. Botts is found.

In the meantime, our chances of making a sale to Mr. Schumaker are diminishing. This afternoon the Mountain Goat Tractor Company had one of their machines out at the sawmill and put on a demonstration. The Mountain Goat is of course far inferior to the Earthworm. But it did fairly good work, and Mr. Schumaker told me that he was so much pleased with it that he had practically decided to buy. I urged him to postpone his decision until we had a chance to show him what our machine could do. He finally agreed to wait until the end of this week. If we cannot give him a demonstration by Saturday afternoon, he will go ahead and buy a Mountain Goat tractor. As this is Wednesday that gives us only three days. Therefore, it is of the utmost importance that you rush another tractor here with all possible speed.

If we sell Mr. Schumaker one tractor this time, and if he finds that it works out well, he may buy fifteen or twenty machines later in the year. He has plenty of work for that many tractors, and plenty of money to buy them.

This is a very important deal. Rush that tractor along. And try to send a man that won't get lost in the woods.

Very truly,
DEANE SUPPLY COMPANY,
John Deane, Pres.

TELEGRAM
SAN FRANCISCO CALIF 8 15A APR 15 1926
JOHN DEANE
MERCEDILLO CALIF

YOUR WIRE RECEIVED STOP WILL SHIP YOU TRACTOR NEXT WEEK STOP WIRE IF ANY FURTHER NEWS FROM BOTTS.

                  FARMERS FRIEND TRACTOR COMPANY

TELEGRAM
MERCEDILLO CALIF 11 02A APR 15 1926
FARMERS FRIEND TRACTOR COMPANY
SAN FRANCISCO CALIF

SHIP TRACTOR AT ONCE RUSH STOP NEXT WEEK WILL BE TOO LATE STOP EXPERIENCED WOODSMEN ARE COMBING THE MOUNTAINS BUT SO FAR NO TRACE OF BOTTS AND TRACTOR

                  JOHN DEANE

TELEGRAM
SAN FRANCISCO CALIF 2 13P APR 15 1926
JOHN DEANE
MERCEDILLO CALIF
AT PRESENT NO TRACTOR AVAILABLE HERE FOR SHIPMENT STOP WE ARE WIRING OUR DEALER IN BAKERSFIELD TO SHIP YOU A SIXTY HORSE POWER EARTHWORM IF POSSIBLE AT ONCE

                  FARMERS FRIEND TRACTOR COMPANY

TELEGRAM
MERCEDILLO CALIF 4 30P APR 15 1926
FARMERS FRIEND TRACTOR COMPANY
SAN FRANCISCO CALIF

STILL NO NEWS OF BOTTS OR TRACTOR STOP PLEASE
DO EVERYTHING POSSIBLE TO EXPEDITE SHIPMENT OF
OTHER TRACTOR AS WE ARE SURE IN A BAD MESS

JOHN DEANE

Farmers' Friend Tractor Company
Salesman's Daily Report

Date: April 15, 1926.
Written from: Glacier Camp, Yosemite Valley, California.
Written by: Alexander Botts, Salesman.
Subject: The Beauties of the Yosemite National Park.

It gives me the greatest pleasure to report that everything is going fine. Since my last report, I have made a slight change in my plans, and I am now spending a few days in the Yosemite Valley. As I am very busy, and as it would take a long time to explain everything, I will merely state that things are coming along great.

If you have never been to the Yosemite Valley, you should by all means take a trip up here. Nothing could be more imposing than El Capitan, unless it be Glacier Point. Nothing could be more inspiring than Yosemite Falls, unless it be the Bridal Veil Falls. And never have I seen anything as beautiful as Mirror Lake. We were both so affected by the beauty of the moon and the stars shining down on this body of water that we stayed out until after midnight last night. Truly this is a paradise on earth.

Most sincerely,
Alexander Botts,
Salesman.

TELEGRAM
SAN FRANCISCO CALIF 9 10A APR 16 1926
ALEXANDER BOTTS
GLACIER CAMP
YOSEMITE VALLEY CAMP

GET BACK TO SCHUMAKERS SAWMILL AT ONCE AND
PUT ON DEMONSTRATION STOP YOUR ABSENCE HAS
SERIOUSLY ENDANGERED SALE TO SCHUMAKER STOP
WIRE THIS OFFICE EXPLANATION OF YOUR CONDUCT
STOP WHAT ARE YOU DOING IN YOSEMITE STOP WHERE
IS THAT TRACTOR STOP HAVE YOU GONE CRAZY

        FARMERS FRIEND TRACTOR COMPANY

TELEGRAM
SAN FRANCISCO CALIF 9 16A APR 16 1926
JOHN DEANE
MERCEDILLO CALIF

WE HAVE LOCATED BOTTS IN YOSEMITE VALLEY
STOP HAVE ORDERED HIM TO RETURN AT ONCE TO
SCHUMAKERS SAWMILL AND PUT ON DEMONSTRATION
STOP DEALER IN BAKERSFIELD WIRES HE HAS NO
TRACTOR ON HAND AVAILABLE FOR SHIPMENT

        FARMERS FRIEND TRACTOR COMPANY

Farmers' Friend Tractor Company
Salesman's Daily Report

Date: April 16, 1926.
Written from: Glacier Camp, Yosemite Valley, California.
Written by: Alexander Botts, Salesman.
Subject: Reply to Your Wire.

I am very much pleased to report that I certainly am going good. Never in my life have I felt more satisfied with myself. And when I tell you everything that I have been doing, you will have to agree that I am getting better and better all the time.

I received your telegram late this afternoon on my return from a long tramp through the woods. There is nothing like a little good strenuous outdoor exercise in congenial company to set a man up and make him feel glad he is alive. In your telegram you asked a number of questions and requested that I wire you an explanation of my conduct. As the explanation will be a little long, I have decided to write instead of telegraph. And if I mail this report tonight you will get it tomorrow anyway, and the delay will be of small importance. Your telegram sounded as if you were worried about me, but as soon as I explain matters you will see that everything is going wonderfully.

In my report of Tuesday, April 13, I related how I had arrived at the sawmill with the tractor and had arranged for a demonstration on Wednesday afternoon. On Wednesday morning I sat around for an hour or two after breakfast, and then decided that I would take the tractor out for a little spin to make sure that it was in proper running order for the afternoon demonstration.

I drove along the private road which goes from the sawmill through about five miles of woods to the main state road. The machine ran perfectly, and I had covered about three or four miles when I noticed a large and handsome sedan coming up the road toward me. I pulled over to one side, and when the sedan got opposite me it stopped and I was delighted to observe that it was driven by Miss Mildred Deane. As I think I mentioned before, she is a very attractive girl. And I was naturally very much pleased and flattered that she had not forgotten her promise to stop off at the sawmill and assist me in this sale to Mr. Schumaker.

She at once introduced me to the other people in the car. There was a rather uninteresting looking elderly lady who was called Aunt Eunice. She was Miss Deane's aunt. There was a young man by the name of Ernest

Clarkson. And there was Miss Sally Schumaker, the daughter of the lumber king, who was a small dark-haired girl, fairly good-looking, but nothing to compare with Miss Mildred Deane. Perhaps I forgot to say that Miss Deane is most remarkable.

Furthermore, she has a most persuasive way about her, as was at once apparent by the way she began asking me to teach her how to drive the tractor.

"It can't be done," I said, very decidedly. "A girl has no business driving a big heavy machine like a tractor."

"What you need to do," she said, "is to calm yourself down and talk sense. If you teach me to run that thing, I can drive it in the demonstration, and when Mr. Schumaker sees that your machine is so simple that even a child like myself can handle it he will be very favorably impressed. Come on, let's go."

She got out of the car, climbed up onto the tractor seat, and smiled at me most pleasantly. I at once realized that if she would only smile at old man Schumaker in that way, he would be completely lost, and would buy anything. I am rather impervious myself to the charms of the opposite sex, but my logical mind was impressed by the young lady's reasoning. If she were to drive the tractor in the demonstration, it could not help but make it more impressive. Furthermore, it was clearly my duty as a salesman to be polite and agreeable to the best friend of the daughter of my perspective customer, Mr. Schumaker.

"All right," I said, "move over here and take hold of these handle bars."

Miss Deane learned very fast. She was already a good automobile driver, and she is so intelligent that she got the knack of handling the tractor almost at once.

Note: On former occasions, when teaching a beginner how to operate a tractor, I have usually sat beside him and given him verbal directions as to what he should do. I have now discovered, however, that things go much better if teacher and pupil both squeeze into the driver's seat and both take hold of the handle bars.

After running up and down the road for a half an hour or so Miss Deane had acquired perfect control of the tractor. We drove back to the parked sedan and stopped. The three people in the car seemed to be getting a little impatient.

"I think, Mildred, that we had better be going," said Aunt Eunice, in a rather whining tone of voice. "If we are going to reach the Yosemite in time to see anything this afternoon, we shall have to be moving."

"Right you are," said Miss Deane. She then turned to me. "Alex, old thing," she said, "we'll have to run up to the sawmill and put on that demonstration just as fast as possible."

"But we can't," I said. "Mr. Schumaker is off in the woods somewhere. He won't be back until this afternoon."

"I can't stay that long," said Miss Deane. "I thought your demonstration would be this morning, and I have promised these people to get them to the Yosemite."

"This is terrible," I said. "I was counting on you to help me."

"I'll tell you what we'll do," said Miss Deane. "We'll put off the demonstration, and you can come along with us to the Yosemite. Then when we get back, after two or three days, you will be all rested up and in wonderful shape to give Mr. Schumaker a small sales talk. And I will have plenty of time to drive the tractor around and help you in every way I can."

"But I have arranged to give the demonstration this afternoon," I said. I spoke these words very positively, to let her see that I am a man who is not easily swayed from his purpose.

"Mr. Schumaker has got along all his life so far without a tractor," said Miss Deane, "and it won't hurt him to wait a couple of days more. Besides, you are the very person we need to make this party a success. Sally and Ernest here are secretly engaged or something. Anyway, they won't look at anybody but each other. If you come along there will be four of us to take long walks through the woods and play around together, while Aunt Eunice sits on the porch and tends to her knitting. If you don't come, I will either have to butt in on Sally and Ernest, or take walks all by myself, or sit on the porch with Aunt Eunice and knit."

"You could take walks with Aunt Eunice," I said.

"No," said Miss Deane. "Aunt Eunice has the rheumatism. All she can do is sit on the porch and knit. There is no use arguing about it, Alex. You have got to come. If you don't, you will just ruin our whole trip to the Yosemite; and I know you are too much of a gentleman to want to do anything like that."

"Well," I said, "of course I would hate to spoil the whole party."

"And you can't do any good here anyway," she went on. "It's beginning to rain. If you put on a demonstration in a shower, Mr. Schumaker will get all wet and cold, and he will not be in a pleasant, receptive state of mind."

At this moment a raindrop splashed onto my nose. "You are right," I said. "It is indeed starting to rain; and probably you are correct in thinking that a wet weather demonstration would be a cheerless affair."

"So, you'll come with us?"

"I shall be pleased," I said, "to accept your kind invitation."

As Aunt Eunice seemed in a hurry, I decided to park the tractor in the woods instead of taking the time to drive it back to the mill. I took especial pains to conceal it, because I did not want anyone tampering with it during my absence. By driving up the middle of a brook which crossed the road nearby I avoided making any tracks which curious people might have followed. I finally parked the machine in a clump of bushes between the stream and a large overhanging rock. In this place I was sure that no one would find it in a hundred years.

After hiding the tractor I hurried back to Miss Deane's automobile, and we were soon speeding on our way toward the Yosemite Valley.

I have given this very full account of everything that happened that morning, in order that you may see that I am, now as always, up on my toes every minute, and that I am conducting this Schumaker deal with energy, initiative and sound common sense.

A less intelligent salesman would probably have stayed behind and stupidly insisted on holding the demonstration on the afternoon scheduled.

"There is no use arguing about it, Alex."

But I am happy to say that I had the good sense to change my plans. The results are most encouraging. I avoided giving a demonstration in a dismal rain. And with my usual politeness, courtesy and tact I have obtained the good will of the best friend of the daughter of the man to whom I expect to sell this tractor. When I get back to Mr. Schumaker's sawmill and put on a demonstration and sales talk with the help of Miss Mildred Deane, it is obvious that the sale will go across quickly and smoothly.

Our trip to the Yosemite has been a complete success. We have now spent two full days enjoying the beauties of Nature in this magnificent valley. And I am glad to report that I have been uniformly courteous and polite, and have lost no opportunity for making myself agreeable to Mr. Schumaker's daughter and to her best friend—thus paving the way most effectively for the coming tractor sale.

Just as Miss Deane prophesied, Aunt Eunice has spent practically all the time sitting on the porch of one of the camp houses. And she actually has her knitting with her and is completely absorbed in the production of some sort of a sweater. Seldom have I seen anyone who seemed to have such a completely negative personality.

The other four of us have put in our time rambling about through the woods and enjoying the great outdoors. It was at once apparent that Ernest and Miss Sally Schumaker were very much attached to each other. And for a while Mildred and I were considerably annoyed by their conduct. At frequent intervals as we walked through the woods Sally would become fatigued. She would sit down on a rock or other convenient resting place. At once Ernest would sit down beside her. And before long they would have their arms around each other, and would be putting on a very heavy petting party.

At first this bothered me a good deal. I knew that at the time they were completely oblivious to Mildred and myself. But I was afraid that after a while they might chance to look up and observe the disgust which we could not keep out of our faces. They would then suddenly realize the spectacle they were making of themselves, and they would be so overcome by embarrassment that the pleasure of our woodland excursion would be ruined.

As I have said I was most anxious to do the polite and courteous thing. And I was suddenly reminded of a story I had once heard of a very courteous action in years gone by. The story seemed so apropos that I told it to Mildred at once.

"It seems," I said, "that some sort of a Turkish or East Indian prince was once dining with Kind Edward in London. The prince didn't know anything about table manners in England, and he proceeded to take a drink out of his finger bowl. Now that may be all right in Turkey, or Brooklyn, or North Fork, Iowa. But at Buckingham Palace it is pretty near a felony. Everybody was all in a flutter; they were afraid that the prince would all of a sudden find out what a boner he had pulled and he would be so embarrassed that the whole party would be a flop.

"But the king was a real master of the art of being polite, and he did exactly the right thing. He took a long drink out of his own finger bowl. All the other guests then did the same, and everything was all right."

As I finished this story, I looked over at Sally and Ernest. And I regret to state that they were hugging each other like the hero and heroine in the slow motion pictures.

"Some people," I said, "just don't seem to know how to behave themselves in polite society."

"Ain't it the truth?" said Mildred sadly. "But I suppose it is our duty to be polite, just as King Edward was."

This turned out to be a very good idea. And from then on we were able to keep Ernest and Sally from being embarrassed by pretending to

"Some people," I said, "just don't seem to know how to behave themselves in polite society."

go through the same motions that they were and deceiving them into thinking that we enjoyed the same sort of thing that they did. In justice to myself I may say that we put on a pretty good imitation. At times we were much better than the original.

I have included these events in my report so that you can see how well I am handling this Schumaker proposition. By my gentlemanly and considerate conduct I am building up a tremendous amount of good will, which will be invaluable when the time comes for those people to use their influence in persuading old Mr. Schumaker to buy a tractor.

In your telegram you request that I go back to the sawmill and put on the demonstration tomorrow. That is entirely in accord with our plans. We all—with the possible exception of Aunt Eunice—hate to tear ourselves away, but we have nevertheless made arrangements to leave this delightful spot tomorrow morning. We will reach the sawmill by noon tomorrow, and after selling Mr. Schumaker the tractor we expect to continue to Mercedillo tomorrow night. From now on, therefore, you should address me in care of the Deane Supply Company, Mercedillo.

I have spent more time than I wanted to on this report, but I felt it was my duty to let you know all of the facts. You do not need to worry any more. Everything is going beautifully.

Very sincerely.
ALEXANDER BOTTS,
Salesman.

TELEGRAM

MERCEDILLO CALIF 8 16P APR 17 1926
FARMERS FRIEND TRACTOR COMPANY
HARVESTER BUILDING
SAN FRANCISCO CALIF

YOUR WIRE OF YESTERDAY STATED THAT BOTTS WOULD BE ON HAND TODAY SATURDAY TO PUT ON DEMONSTRATION AT SAWMILL STOP WE WAITED ALL DAY BUT BE NEVER SHOWED UP STOP SCHUMAKER HAS BOUGHT MOUNTAIN GOAT TRACTOR STOP THIS LEAVES US OUT IN THE COLD COMPLETELY STOP YOU ARE THE SLOPPIEST AND MOST IDIOTIC PEOPLE TO DO BUSINESS

WITH I EVER HEARD OF STOP IF YOU CANT COOPERATE WITH ME BETTER THAN THIS I WILL CANCEL MY CONTRACT WITH YOU

            JOHN DEANE

TELEGRAM

SAN FRANCISCO CALIF 9 02A APR 19 1926
JOHN DEANE
MERCEDILLO CALIF

YOUR WIRE OF LAST SATURDAY NIGHT RECEIVED THIS MONDAY MORNING STOP REGRET EXCEEDINGLY THAT THINGS HAVE GONE WRONG STOP I WILL GO TO MERCEDILLO MYSELF THE LATTER PART OF THIS WEEK TO TALK THINGS OVER WITH YOU

            J D WHITCOMB
         WESTERN SALES MANAGER
      FARMERS FRIEND TRACTOR COMPANY

TELEGRAM

SAN FRANCISCO CALIF APR 19
ALEXANDER BOTTTS
CARE OF DEANE SUPPLY COMPANY
MERCEDILLO CALIF
HOLD UNTIL CALLED FOR

WE REGRET THAT RECENT EVENTS MAKE IT IMPOSSIBLE FOR US TO RETAIN YOU IN OUR EMPLOYMENT STOP KINDLY FORWARD US YOUR FINAL EXPENSE ACCOUNT WITH BALANCE OF ADVANCE EXPENSE MONEY AND WE WILL MAIL YOU OUR CHECK FOR SALARY DUE YOU TO DATE

            J D WHITCOMB
         WESTERN SALES MANAGER
      FARMERS FRIEND TRACTOR COMPANY

Farmers' Friend Tractor Company
Salesman's Daily Report

Date: April 19, 1926.
Written from: Schumaker Sawmill.
Written by: Alexander Botts, Salesman.
Subject: Everything Going Fine.

This is just a short report to let you know that all is well. We haven't sold the tractor as yet, but I am not worried at all. In fact, I consider myself the luckiest and most fortunate man in the whole world.

In my last report I told you that we expected to leave the Yosemite on Saturday morning. Later, however, we changed our plans. It was after supper on Friday night that Mildred suddenly announced her desire to stay in the Yosemite over Sunday. At first I was opposed to this. But naturally I would not want to be impolite, so I decided to remain also. And later events proved that, as usual, I was right in my decision.

It was early Saturday morning that I conceived one of the most brilliant ideas that has ever come to me in my life. It suddenly occurred to me that as long as Mildred was going to be such a great help to me in this Schumaker tractor sale, I ought to try to arrange matters so that she could help me in all future affairs of this kind.

As soon as possible I took Mildred for a walk in the woods and asked her what she thought about the matter. And I was delighted to find that she grasped my point of view at once. As I have said before she is a most remarkable person. She is the only girl I ever met who has the intellect and mental power to understand exactly what I am talking about at all times. In fact, she even seems to understand what I am thinking about, whether I say anything or not. And in less than five minutes we had not only decided to get married but to get married right away. We found Ernest and Sally, and told them they ought to do the same—thus making it a double wedding.

At first Sally objected. "I don't dare," she said.

"Why not?" asked Mildred. "You want to get married, don't you?"

"Yes, but I'm afraid father might be mad. I don't think he approves of Ernest any too much."

"That's the very reason why you should do it now," said Mildred. "Suppose your father does disapprove of Ernest? My father thinks Alex is less than nothing. What if you are afraid your father might get mad?

I am absolutely sure that when my father hears about this, he'll go off like a whole store full of fireworks. That's why we have to get it over quick. The longer we delay, the longer it will string out the shouting and hollering and arguments. But when the thing is done, there will be one grand explosion and then things will calm down. The world accepts an accomplished fact."

Naturally, Ernest and Sally were both much impressed by this masterful argument, and finally they agreed to take the plunge.

There seemed to be a lot of red tape. We had to drive around in the car and hunt up the authorities and sign a lot of papers and get licenses and everything. Then we had to find a minister. So it wasn't until early this morning that the double wedding took place.

So far, we haven't told any of the friends and relatives—not even Aunt Eunice. Mildred did not want to do anything to disturb the old lady at that particular time. You see, Aunt Eunice had almost finished the sweater upon which she had been knitting. And Mildred felt that any unexpected news might tend to make her nervous. Her mind might start wandering from her work; she would probably start dropping stitches, and the entire sweater might, in consequence, be ruined.

Right after the ceremony the four of us returned to the camp, loaded Aunt Eunice and the baggage into the car, and drove down here to Mr. Schumaker's sawmill, arriving about eleven this morning. I found the tractor just where I had left it, and brought it in to the mill.

We learned that Mr. Schumaker—following what seems to be his usual custom—was out in the woods somewhere and was not expected back until about three this afternoon. It is now half-past two. We have had luncheon and I have been improving my time while waiting by writing this report. Mr. Schumaker should soon be here, and when he arrives we will surprise him with one of the finest tractor demonstrations he has ever seen.

Later. Mercedillo, California, 9 p.m.

Much of interest has happened since I wrote the first part of this report.

Mr. Schumaker got back to the mill, as was expected about three this afternoon. At first he did not want to see any demonstrations, but Mildred and his daughter Sally and I talked to him so convincingly and persuasively that he finally let us show him what we could do.

And we showed him plenty. Mildred did most of the driving, and the way that girl made the old tractor snake logs up and down hill, over rocks and through brooks and everywhere was a marvel to see. When the exhibition was over, Mr. Schumaker told us he was very much impressed. It appears that the poor fool had had no more sense than to buy himself one of those Mountain Goat tractors a couple of days before.

"But I never would have done it," he admitted, "if I had known how much better your machine is."

As soon as I heard him say this I started in and told him he ought to buy one of our machines also. But right away Mildred interrupted.

"One Earthworm isn't enough," she said, very positively. "With the number of logs you have to move, Mr. Schumaker, you need at least twelve machines. And as long as father isn't here today to help Mr. Botts, I am going to talk this thing out with you."

And she went ahead and gave a sales talk that was as logical and convincing as anything I could have done myself. And it was a whole lot more pleasing and interesting. Sally Schumaker told her father that the proposition looked good to her. Fortunately Ernest had the good sense to keep out of sight, while Aunt Eunice, as was to be expected, kept on with her knitting.

Finally the old man came around. He did not take twelve machines, but he did sign up for eight Sixty Horse Power Earthworms, and his order will be sent in to you through the office of the Deane Supply Company. He retained the demonstration machines and wants the other seven shipped at once.

As soon as the papers were signed, Mildred and I placed Aunt Eunice in the sedan, climbed in ourselves, said goodbye, and started for Mercedillo to break the glad news to father—leaving Ernest and Sally at the mill to spring their own pleasant little surprise on Mr. Schumaker.

Note: I forgot to say that we had decided not to spill the news of the happy marriages until after we had sold Mr. Schumaker the tractors. We thought that it was wiser not to start any sort of a family discussion which might have distracted the old gentleman's attention from the business in hand.

On the way to Mercedillo Aunt Eunice told us that she had finished the sweater and Mildred told her that we had been married. At once Aunt Eunice was all in a flutter with excitement, and I realized that Mildred had been perfectly right in letting her finish the sweater before telling her this news. After twittering and bubbling somewhat incoherently for a while, Aunt Eunice settled down to one remark: "What will your father say?" She kept repeating: "What will your father say?"

"I haven't any idea what he will say," remarked Mildred pleasantly, "but it's sure to be good. If you want to get an earful, aunty, you had better stick around when I tell him."

Upon reaching Mercedillo we drove directly to Mr. Deane's residence.

"I want you to leave everything to me, Alex," said Mildred as we entered the front door. "I want to handle this business myself."

"Just as you say, my dear," I said.

So I remained in the front hall while Aunt Eunice fled upstairs and Mildred went into the living room to let father in on our little secret. I didn't hear what she said to him, but she must have told him all about it right away, because he began shouting and hollering in what I can only describe as a very loud and vulgar way. And after he had told Mildred what he thought of her, he came out in the hall and started to explain what he thought of me.

"This is an outrage!" he said. "An outrage! What do you mean, you dirty little city slicker, by presuming to marry my only daughter? I'll have it annulled! And what's more I'll wring your scrawny little neck!"

"At-a-boy, father!" said Mildred, who was behind him. "You're going fine! The louder you yell, and the more you work yourself up, the sooner you'll get through and quiet down so we can talk sense."

At these remarks Mr. Deane went into even a greater frenzy than before, pacing up and down, waving his arms around and positively foaming at the mouth. I do not remember exactly what he said, but I gathered that he was distinctly displeased both with his daughter and with myself, and that he wanted me to get out of the house before he kicked me out, and never to darken his door with my obnoxious presence again.

As may be imagined, all this placed me in a somewhat embarrassing position. But I am proud to say that I retained my composure, and did not demean myself by talking back to him or by arguing with him in any way. I merely told him, as politely and tactfully as I could, that he was making a fool of himself, that I did not intend to leave, and that I sincerely hoped he would try to kick me out, so that I would have a good excuse for knocking his block off. My calmness, instead of quieting him down, only seemed to infuriate him more. But for some reason or other he made no attempt to eject me by force. And finally, just as Mildred had predicted, he got himself so tired by yelling that he began to subside through pure weariness. As soon as he was thoroughly talked out. Mildred led him gently back into the living room. I followed along, and we all sat down and talked things over. Mildred summed up the situation in her usual clear and convincing manner.

Mr. Deane went into even a greater frenzy than before.

"We're married," she said, "and we're going to stay married. Alex and I are both more than twenty-one. We both know what we want. So there is no use arguing about it any more. The only thing we need to discuss is the future."

"I can't stand it," said the old man weakly, "to see you go off with this man and leave me all alone. You're the only daughter I have."

At this Mildred smiled at him very affectionately. "You're a pretty good old guy, dad, even if you have a vile temper," she said. "I'm very fond of you, and I'm sure you're going to do the proper thing. If you want to keep your daughter in town, you'll have to keep your son-in-law in town. And the only way to keep your son-in-law in town is to give him a job."

"Yeah?" said Mr. Deane.

"Absolutely!" said Mildred. She then talked along in her usual convincing way, explaining exactly what she wanted. And after a while Mr. Deane came around and agreed to everything.

It has all been decided. I am to be vice president of the Deane Supply Company, in full charge of tractor sales. In this position I will have an opportunity to make an even greater use of my unusual selling ability than I did as a salesman employed directly by the Farmers' Friend Tractor Company. And I am glad to know that you seem to approve the change I am making.

Mr. Deane has just handed me your telegram in which you say, "We regret that recent events make it impossible for us to retain you in our employ." This telegram is worded in such a way that when I first read it I thought it sounded as if you were trying to fire me. On second thought I realized that this could not be so, because of course you could not conceivably have any reason for wanting to get rid of me. I can only assume that you must have guessed the course which events were taking, and wished to save me the pain of resigning.

I wish to assure you that my regret at leaving the employ of the Farmers' Friend Tractor Company is as great as your regret in losing me. And I am sure that my new relation to you as dealer will be in every way as satisfactory as our former association.

I am sure that I am going to be a far, far better tractor salesman than ever before. Mrs. Botts is sure to be invaluable as an assistant. I cannot get over my admiration for the way she is able to manage such men as Mr. Schumaker and her father. They seem to do exactly what she tells them. In fact, I seem to be the only man she has ever said that she cannot boss around. Perhaps that is why she likes me.

In closing this, my last report to the Farmers' Friend Tractor Company, I wish to say that Mrs. Botts and I are going to buy a very comfortable little house on the edge of town. I had thought of getting an apartment down near the office. But she didn't like the idea, so I have decided on this house farther away.

Very sincerely,
ALEXANDER BOTTS.

ABOUT WILLIAM HAZLETT UPSON

WILLIAM HAZLETT UPSON was born in Glen Ridge, New Jersey, September 26, 1891. His father was a Wall Street lawyer, his mother a doctor of medicine, and most of the rest of the family doctors, lawyers, college professors, and engineers. To be different, Bill Upson became a farmer—but found the job involved too much hard work. He escaped from the farm by enlisting in the field artillery in World War I.

After the war, he worked from 1919 to 1924 as a service mechanic and troubleshooter for the Caterpillar Tractor Company. "My main job," he said, "was to travel around the country trying to make the tractors do what the salesman had said they would. In this way I learned more about salesmen than they know about themselves." He also became very fond of salesmen. He admitted they are crazy, but maintained they are splendid people with delightful personalities.

In 1923, Bill began writing short stories. In 1927, he created the character Alexander Botts, who has appeared in over a hundred *Saturday Evening Post* stories.

Bill was married to Marjory Alexander Wright. For many years, their home was in Middlebury, Vermont. They had a son, Job Wright Upson, a daughter, Polly (Mrs. Claude A. Brown), and a dog, Shelley. Upson spent his last days in Middlebury with his family, and passed away on February 5, 1975, at the age of 83.

ABOUT ALEXANDER BOTTS*

Requests for biographical material have come from Botts fans, particularly the younger ones who missed many of the earlier stories. From the files of the EARTHWORM CITY IRREGULARS—the national Botts fan club— come the following notes:

ALEXANDER BOTTS was born in Smedleytown, Iowa, on March 15, 1892, the son of a prosperous farmer. He finished high school there, then embarked on a series of jobs—none of them quite worthy of his mettle. In these early days the largest piece of machinery he sold was the Excelsior Peerless Self-Adjusting Automatic Safety Razor Blade Sharpener. He became interested in heavy machinery in 1918 while serving in France as a cook with the motorized field artillery. In March 1920, he was hired as a salesman by the Farmers' Friend Tractor Company, which later became the Earthworm Tractor Company.

On April 12, 1926, he met Miss Mildred Deane, the attractive daughter of an Earthworm dealer in Mercedillo, California. Seven days later they were married. Mildred, later nicknamed Gadget, had attended the language schools at Middlebury College (Vermont) and acted as interpreter for her husband when he was sent to Europe in 1928 to open new tractor outlets there.

Mr. and Mrs. Botts returned from Europe in early 1929 to await the birth of Alexander Botts, Jr., who arrived in February along with a twin sister, Little Gadget. Mr. Botts now has been a grandfather for some years.

The adventures of Botts have been appearing in *The Saturday Evening Post* since 1927, over a hundred stories in all. They have been collected into eight books, most of them now out of print. Your local bookstore probably can locate used copies of the books, or, if not, the EARTHWORM CITY IRREGULARS office usually has copies on hand. Write to R. D. Blair, 38 Main Street, Middlebury, Vermont, for information, and—if you're a true-blue Botts fan—ask about joining the IRREGULARS.

This biography was written by William Hazlett Upson and published in the 1963 book Original Letters of Alexander Botts *by Vermont Books: Publishers in Middlebury. While the EARTHWORM CITY IRREGULARS are no longer meeting, Upson's local bookstore— The Vermont Book Shop—which opened in 1949 is still open at this address today.*

We are pleased to have presented the first installments of the Alexander Botts saga in their entirety, along with the illustrations that accompanied them in the *Saturday Evening Post*.

We intend to present the full collection of over 100 short stories, including some that were published outside of the *Post*. To see if Botts returns to the Farmers' Friend Tractor Company in the next volume, visit octanepress.com or anywhere books are sold.

Follow us on Facebook, Instagram, Twitter, or the latest, new-fangled social media platform to learn more!